MW00717277

Testing of Concrete in Structures

Testing of Concrete in Structures

Third Edition

J.H. BUNGEY
Professor of Civil Engineering
University of Liverpool

and

S.G. MILLARD
Senior Lecturer in Civil Engineering
University of Liverpool

BLACKIE ACADEMIC & PROFESSIONAL
An Imprint of Chapman & Hall

London · Glasgow · Weinheim · New York · Tokyo · Melbourne · Madras

Published by
Blackie Academic & Professional, an imprint of Chapman & Hall,
Wester Cleddens Road, Bishopbriggs, Glasgow G64 2NZ

Chapman & Hall, 2–6 Boundary Row, London SE1 8HN, UK

Blackie Academic & Professional, Wester Cleddens Road, Bishopbriggs, Glasgow G64 2NZ, UK

Chapman & Hall GmbH, Pappelallee 3, 69469 Weinheim, Germany

Chapman & Hall USA, Fourth Floor, 115 Fifth Avenue, New York, NY 10003, USA

Chapman & Hall Japan, ITP-Japan, Kyowa Building, 3F, 2-2-1 Hirakawacho, Chiyoda-ku, Tokyo 102, Japan

DA Book (Aust.) Pty Ltd, 648 Whitehorse Road, Mitcham 3132, Victoria, Australia

Chapman & Hall India, R. Seshadri, 32 Second Main Road, CIT East, Madras 600 035, India

First edition 1982
Second edition 1989
Reprinted 1994
This edition 1996

© 1996 Chapman & Hall

Typeset in 10/12pt Times by AFS Image Setters Ltd, Glasgow

Printed in Great Britain by The University Press, Cambridge

ISBN 0 7514 0241 9

A catalogue record for this book is available from the British Library
Library of Congress Catalog Card Number: 95-80537

∞ Printed on acid-free text paper, manufactured in accordance with ANSI/NISO Z39.48-1992 (Permanence of Paper)

Preface

Interest in testing of hardened concrete in-situ has increased considerably since the 1960s, and significant advances have been made in techniques, equipment, and methods of application since publication of the first edition of this book in 1982. This has largely been a result of the growing number of concrete structures, especially those of recent origin, that have been showing signs of deterioration. Changes in cement manufacture, increased use of cement replacements and admixtures, and a decline in standards of workmanship and construction supervision have all been blamed. Particular attention has thus been paid to development of test methods which are related to durability performance and integrity. There is also an increasing awareness of the shortcomings of control or compliance tests which require a 28-day wait before results are available. Even then, such tests reflect only the adequacy of the material supplied rather than overall construction standards.

In each case the need for in-situ measurements is clear, but to many engineers the features, and especially the limitations, of available test methods are unknown and consequently left to 'experts'. Although it is essential that the tests should be performed and interpreted by experienced specialists, many difficulties arise both at the planning and interpretation stages because of a lack of common understanding. A great deal of time, effort and money can be wasted on unsuitable or badly planned testing, leading to inconclusive results which then become the subject of heated debate.

The principal aim of this book is to provide an overview of the subject for non-specialist engineers who are responsible for the planning of test programmes. The scope is wide in order to cover comprehensively as many aspects as possible of the testing of hardened concrete in structures. The tests, however, are treated in sufficient depth to create a detailed awareness of procedures, scope and limitations, and to enable meaningful discussions with specialists about specific methods. Carefully selected references are also included for the benefit of those who wish to study particular methods in greater detail. The information and data contained in the book have been gathered from a wide variety of international sources. In addition to established methods, new techniques which show potential for future development are outlined, although in many cases the application of these to concrete is still at an early stage and of limited practical value at present. Emphasis has been placed on the reliability and limitations of the various techniques described, and the interpretation of results is discussed from the point of view both of specification compliance and application to design calculations. A number of illustrative examples have been included with this in mind.

In preparing this third edition the original author has been joined by his colleague Dr Steve Millard and the opportunity has been taken to reflect trends in equipment and procedures which have developed over the past six years. These include a general move to automate test methods, particularly in terms of data collection, storage and presentation. It is important to recognize that this does not imply increased accuracy. Interest in the application of statistical methods to interpretation of strength test results has grown, especially in the USA, and details have been incorporated into Chapter 1 and illustrated by an additional example in Appendix A.

The growing importance of performance monitoring is reflected in Chapter 6, whilst Chapter 7 which deals with durability has a good deal of new material relating to corrosion assessment and surface zone permeability measurement techniques. Both of these areas have seen significant developments of apparatus and procedures since the second edition was prepared. The coverage of sub-surface radar, which has now become an established technique, is similarly increased in Chapter 8, whilst many other developments have also been incorporated throughout the book. References to Standards have been updated and a significant number of recent new references have been added.

The basic testing techniques will be similar in all parts of the world, although national standards may introduce minor procedural variations and units will of course differ. This book has been based on the SI units currently in use in Britain, and where reference to Codes of Practice has been necessary, emphasis is placed on the current recommendations of British Standards. The most recent versions of standards should always be consulted, since recommendations will inevitably be modified from time to time.

We are very grateful to many engineers worldwide for discussions in which they have provided valuable advice and guidance. Particular thanks are due to members of the former BSI Subcommittee CAB/4/2 (Non-destructive Testing of Concrete) for the stimulation provided by their contributions to meetings of that subcommittee and to our colleague at Liverpool, Mr R.G. Tickell for his assistance with statistical material. Our thanks are also extended to Mr M. Grantham of M.G Associates for his advice relating to the updating of Chapter 9 on Chemical Analysis. Photographs have been kindly provided by Tysons Plc, Professor F. Sawko, CNS Electronics Ltd, Steinweg UK Ltd, Hammond Concrete Services, Agema Infrared Systems, Protimeter Plc, Dr A. Poulsen and Germann Aps and these are gratefully acknowledged. Thanks are also due to Ms M.A. Revell for typing the original manuscript, to Ms A. Ventress for typing new material associated with this edition and Mrs B. Cotgreave for preparation of the diagrams.

J.H.B.
S.G.M.

Contents

1 Planning and interpretation of in-situ testing

A great deal of time, effort and expense can be wasted on in-situ testing unless the aims of the investigation are clearly established at the outset. These will affect the choice of test method, the extent and location of the tests and the way in which the results are handled — inappropriate or misleading test results are often obtained as a result of a genuine lack of knowledge or understanding of the procedures involved. If future disputes over results are to be avoided, liaison of all parties involved is essential at an early stage in the formulation of a test programme. Engineering judgement is inevitably required when intepreting results, but the uncertainties can often be minimized by careful planning of the test programme.

A full awareness of the range of tests available, and in particular their limitations and the accuracies that can be achieved, is important if disappointment and disillusion is to be avoided. Some methods appear to be very simple, but all are subject to complex influences and the use of skilled operators and an appropriately experienced engineer are vital.

In-situ testing of existing structures is seldom cheap, since complex access arrangements are often necessary and procedures may be time-consuming. Ideally a programme should evolve sequentially, in the light of results obtained, to provide the maximum amount of worthwhile information with minimum cost and disruption. This approach, which requires ongoing interpretation, will also facilitate changes of objectives which may arise during the course of an investigation.

1.1 Aims of in-situ testing

Three basic categories of concrete testing may be identified.

(i) *Control testing* is normally carried out by the contractor or concrete producer to indicate adjustments necessary to ensure an acceptable supplied material.
(ii) *Compliance testing* is performed by, or for, the engineer according to an agreed plan, to judge compliance with the specification.
(iii) *Secondary testing* is carried out on hardened concrete either in, or extracted from, the structure. This may be required in situations where there is doubt about the reliability of control and compliance results or they are unavailable or inappropriate, as in an old, damaged, or deteriorating structure. All testing which is not planned before

construction will be in this category, although long-term monitoring is also included.

Control and compliance tests have traditionally been performed on 'standard' hardened specimens made from samples of the same concrete as used in a structure; it is less common to test fresh concrete. There are also instances in which in-situ tests on the hardened concrete may be used for this purpose. This is most common in the precasting industry for checking the quality of standardized units, and the results can be used to monitor the uniformity of units produced as well as their relationship to some pre-established minimum acceptable value. There is, generally, an increasing awareness amongst engineers that 'standard' specimens, although notionally of the same material, may misrepresent the true quality of concrete actually in a structure. This is due to a variety of causes, including non-uniform supply of material and differences of compaction, curing and general workmanship, which may have a significant effect on future durability. As a result, a trend towards in-situ compliance testing, using methods which are either non-destructive or cause only very limited damage, is emerging, particularly in North America and Scandinavia. Such tests are most commonly used as a back-up for conventional testing, although there are notable instances such as the Storebaelt project where they have played a major role (1). They offer the advantage of early warning of suspect strength, as well as the detection of defects such as inadequate cover, high surface permeability, voids, honeycombing or use of incorrect materials which may otherwise be unknown but lead to long-term durability problems. Testing of the integrity of repairs is another important and growing area of application.

The principal usage of in-situ tests is nevertheless as *secondary* testing, which may be necessary for a wide variety of reasons. These fall into two basic categories.

1.1.1 *Compliance with specification*
The most common example is where additional evidence is required in contractual disputes following non-compliance of standard specimens. Other instances involve retrospective checking following deterioration of the structure, and will generally then be related to apportionment of blame in legal actions. Strength requirements form an important part of most specifications, and the engineer must select the most approriate methods of assessing the in-situ strength on a representative basis, with full knowledge of the likely variations to be expected within various structural members (as discussed in section 1.5). The results should be interpreted to determine in-situ variability as well as strength, but a major difficulty arises in relating measured in-situ strength to anticipated corresponding 'standard' specimen strength at a specific but different age. Borderline cases may thus be difficult to prove conclusively. This problem is examined in detail in section 1.5.2.

Minimum cement content will usually be specified to satisfy durability requirements, and chemical or petrographic tests may be necessary to confirm compliance. Similar tests may also be required to check for the presence of forbidden admixtures, contamination of concrete constituents (e.g. chlorides in sea-dredged aggregates) or entrained air, and to verify cement content following deterioration. Poor workmanship is often the principal cause of durability problems, and tests may also be aimed at demonstrating inadequate cover or compaction, incorrect reinforcement quantities or location, or poor quality of curing or specialist processes such as grouting of post-tensioned construction.

1.1.2 *Assessment of in-situ quality and integrity*

This is primarily concerned with the current adequacy of the existing structure and its future performance. Routine maintenance needs of concrete structures are now well established, and increasingly utilize in-situ testing to assist 'lifetime predictions' (2, 3). It is important to distinguish between the need to assess the properties of the material, and the performance of a structural member as a whole. The need for testing may arise from a variety of causes, which include

  (i) Proposed change of usage or extension of a structure
 (ii) Acceptability of a structure for purchace or insurance
(iii) Assessment of structural integrity or safety following material deterioration, or structural damage such as caused by fire, blast, fatigue or overload
 (iv) Serviceability or adequacy of members known or suspected to contain material which does not meet specifications, or with design faults
  (v) Assessment of cause and extent of deterioration as a preliminary to the design of repair or remedial schemes
 (vi) Assessment of the quality or integrity of applied repairs
(vii) Monitoring of strength development in relation to formwork stripping, curing, prestressing or load application
(viii) Monitoring long-term changes in materials properties and structural performance.

Although in specialized structures, features such as density or permeability may be relevant, generally it is either the in-situ strength or durability performance that is regarded as the most important criterion. Where repairs are to be applied using a different material from the 'parent' concrete, it may be desirable to measure the elastic modulus to determine if strain incompatibilities under subsequent loading may lead to a premature failure of the repair. A knowledge of elastic modulus may also be useful when interpreting the results of load tests. For strength monitoring during construction, it will normally only be necessary to compare test results with limits established by trials at the start of the contract, but in other situations a prediction of actual concrete strength is required to incorporate into calculations of member strength. Where calculations are to be based on measured in-situ strength, careful attention mut be paid to the numbers and location of tests and the validity of the safety factors adopted, and this problem is discussed in section 1.6.

Durability assessments will concentrate upon identifying the presence of internal voids or cracking, materials likely to cause disruption of the concrete (e.g. sulphates or alkali-reactive aggregates), and the extent or risk of reinforcement corrosion. Carbonation depths, chloride concentrations, cover thicknesses and surface zone resistivity and permeability will be key factors relating to corrosion. Electrochemical activity associated with corrosion can also be measured to assess levels of risk, using passive or perturbative test methods.

Difficulties in obtaining an accurate quantitative estimate of in-situ concrete properties can be considerable: wherever possible the aim of testing should be to compare suspect concrete with similar concrete in other parts of the structure which is known to be satisfactory, or of proven quality.

Investigation of the overall structural performance of a member is frequently the principal aim of in-situ testing, and it should be recognized that in many situations this would be most convincingly demonstrated directly by means of a load test. The confidence attached to the findings of the investigation may then be considerably greater than if member strength predictions are derived indirectly from strength estimates based on in-situ materials tests. Load testing may however be prohibitively expensive or simply not a practical proposition.

1.2   Guidance available from 'standards' and other documents

National standards are available in a number of countries, notably the UK, USA and Scandinavia, detailing procedures for the most firmly established testing methods. Principal British and ASTM standards are listed at the end of this chapter and specific references are also included in the text. ISO standards are also under development in some cases. Details of all methods are otherwise contained in an extensive body of published specialist research papers, journals, conference proceedings and technical reports. References to a key selection of these are provided as appropriate.

General guidance concerning the philosophy of maintenance inspection of existing structures is provided by FIP (4) and also by the Institution of Structural Engineers (5) who consider appraisal processes and methods as well as testing requirements. Advice is also offered on sources of information, reporting, and indentification of defects with their possible causes. Specific guidance on damage classification is proposed by RILEM (6) whilst ACI committee 364 have produced a guide for evaluation of concrete structures prior to rehabilitation (7). Guidance relating to assessment approaches to specialized situations such as high alumina cement concrete (8), fire (9) and bomb-damaged structures (10) is also available. BS 1881: Part 201, *Guide to the use of non-destructive methods of test for hardened concrete* (11), provides outline descriptions of 23 wide-ranging methods, together with guidance on test selection and planning, whilst BS 6089 (12) relates specifically to in-situ strength assessment. Methods and apparatus which are commercially available are constantly changing and developing, but CIRIA Technical Note 143 (13) reviewed those existing in the UK in 1992 whilst Schickert has outlined the situation in Germany in 1994 (14). Carino has also recently reviewed the worldwide historical development of non-destructive testing of concrete from the North American perspective and has identified future prospects (15). As newer methods become established it is likely that further standards and reports will appear. ACI Committee 228 is currently preparing a substantial report reviewing non-destructive methods whilst RILEM Committee 126 is considering in-place strength testing. In the UK, the Concrete Society is also preparing technical reports on reinforcement corrosion assessment and sub-surface radar methods.

1.3 Test methods available

Details of individual methods are given in subsequent chapters and may be classified in a variety of ways. Table 1.1 lists the principal tests in terms of the property under investigation. The range of available tests is large, and there are others which are not included in the table but are described in this book. Visual inspection, assisted where necessary by optical devices, is a valuable assessment technique which must be included in any investigation. There will of course be overlap of usage of some tests between the applications listed (see section 1.4.3), and where a number of options are available considerations of access, damage, cost, time and reliability will be important.

The test methods may also be classified as follows:

Non-destructive methods. Non-destructive testing is generally defined as not impairing the intended performance of the element or member under test, and when applied to concrete is taken to include methods which cause localized surface zone damage. Such tests are commonly described as partially-destructive and many of those listed in Table 1.1 are of this type. All non-destructive methods can be performed directly on the in-situ concrete without removal of a sample, although removal of surface finishes is likely to be necessary.

Methods requiring sample extraction. Samples are most commonly taken in the form of cores drilled from the concrete, which may be used in the laboratory for strength and other physical tests as well as visual, petrographic and chemical analysis. Some chemical tests may be performed on smaller

Table 1.1 Principal test methods

Property under investigation	Test	Equipment type
Corrosion of embedded steel	Half-cell potential	Electrochemical
	Resistivity	Electrical
	Linear polarization resistance	Electrochemical
	A.C. Impedance	Electrochemical
	Cover depth	Electromagnetic
	Carbonation depth	Chemical/microscopic
	Chloride concentration	Chemical/electrical
Concrete quality, durability and deterioration	Surface hardness	Mechanical
	Ultrasonic pulse velocity	Electromechanical
	Radiography	Radioactive
	Radiometry	Radioactive
	Neutron absorption	Radioactive
	Relative humidity	Chemical/electronic
	Permeability	Hydraulic
	Absorption	Hydraulic
	Petrographic	Microscopic
	Sulphate content	Chemical
	Expansion	Mechanical
	Air content	Microscopic
	Cement type and content	Chemical/microscopic
	Abrasion resistance	Mechanical
Concrete strength	Cores	Mechanical
	Pull-out	Mechanical
	Pull-off	Mechanical
	Break-off	Mechanical
	Internal fracture	Mechanical
	Penetration resistance	Mechanical
	Maturity	Chemical/electrical
	Temperature-matched curing	Electrical/electronic
Integrity and performance	Tapping	Mechanical
	Pulse-echo	Mechanical/electronic
	Dynamic response	Mechanical/electronic
	Acoustic emission	Electronic
	Thermoluminescence	Chemical
	Thermography	Infra-red
	Radar	Electromagnetic
	Reinforcement location	Electromagnetic
	Strain or crack measurement	Optical/mechanical/electrical
	Load test	Mechanical/electronic/electrical

drilled powdered samples taken directly from the structure, thus causing substantially less damage, but the risk of sample contamination is increased and precision may be reduced. Making good the sampling damage will be necessary, as with partially-destructive methods.

The nature of the testing equipment ranges from simple inexpensive hand-held devices to complex, expensive, highly specialized items, possibly requiring extensive preparation or safety precautions, which will be used only where no simple alternative exists. Few of the methods give direct quantitative measurement of the desired property, and correlations will often be necessary. Practical limitations, reliability and accuracy vary widely and are discussed in the sections of this book dealing with the various individual methods. Selection of the most appropriate tests within the categories of Table 1.1 is discussed in section 1.4.3 of this chapter.

1.4 Test programme planning

This involves consideration of the most appropriate tests to meet the established aims of the investigation, the extent or number of tests required to reflect the true state of the concrete, and the location of these tests. Investigations have been made into the use of Expert Systems to assist this process but at the present time it seems likely that their application will be largely confined to a training role (16). Visual inspection is an essential feature whatever the aims of the test programme, and will enable the most worthwhile application of the tests which have been summarized in section 1.3. Some typical illustrative examples of test programmes to meet specific requirements are given in Appendix A.

1.4.1 *General sequential approach*
A properly structured programme is essential, with interpretation as an ongoing activity, whatever the cause or nature of an investigation. Figure 1.1 illustrates the stages typically involved (17), which will generally require increasing cost commitment, and the investigation will proceed only as far as is necessary to reach firm relevant conclusions.

1.4.2 *Visual inspection*
This can often provide valuable information to the well-trained eye. Visual features may be related to workmanship, structural serviceability and material deterioration, and it is particularly important that the engineer be able to differentiate between the various types of cracking which may be encountered. Figure 1.2 illustrates a few of these in their typical forms.

Segregation or excessive bleeding at shutter joints may reflect problems with the concrete mix, as might plastic shrinkage cracking, whereas honeycombing may be an indication of low standards of construction workmanship. Lack of structural adequacy may show itself by excessive deflection or flexural cracking, and this may frequently be the reason for an in-situ assessment of a structure. Long-term creep deflections, thermal movements, or structural movements may cause distortion of door frames, cracking of windows, or cracking of a structure or its finishes. Visual comparison of similar members is particularly valuable as a preliminary to testing to determine the extent of the problem in such cases.

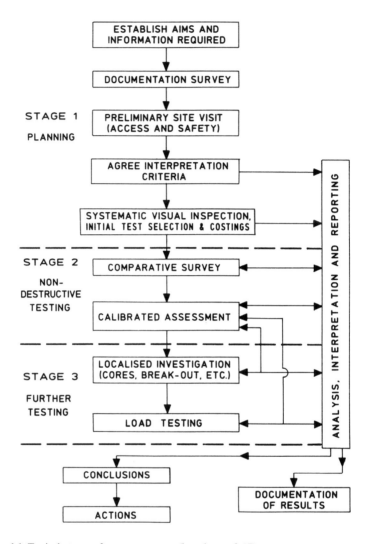

Figure 1.1 Typical stages of test programme (based on ref. 17).

Material deterioration is often indicated by surface cracking and spalling of the concrete, and examination of crack patterns may provide a preliminary indication of the cause. The most common causes are reinforcement corrosion due to inadequate cover or high chloride concentrations, and concrete disruption due to sulphate attack, frost action or alkali-aggregate reactions. As shown in Figure 1.2, reinforcement corrosion is usually indicated by splitting and spalling along the line of bars possibly with rust staining, whereas sulphate attack may produce a random pattern accompanied by a white

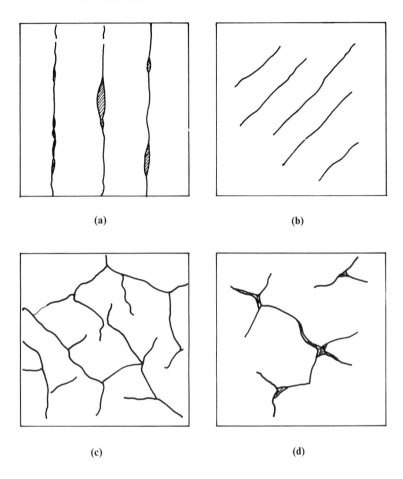

(a)   (b)

(c)   (d)

Figure 1.2 Some typical crack types: (a) reinforcement corrosion; (b) plastic shrinkage; (c) sulphate attack; (d) alkali/aggregate reaction.

deposit leached on the surface. Alkali-aggregate reaction is sometimes (but not necessarily) characterized by a star-shaped crack pattern, and frost attack may give patchy surface spalling and scabbing. Because of similarities it will often be impossible to determine causes by visual inspection alone, but the most appropriate identification tests can be selected on this basis. Careful field documentation is important (18) and Pollock, Kay and Fookes (19) suggest that systematic 'crack mapping' is a valuable diagnostic exercise when determining the causes and progression of deterioration, and they give detailed guidance about the recognition of crack types. Non-structural cracking is described in detail by Concrete Society Technical Report 22 (20), and the symptoms relating to the most common sources of deterioration are

Table 1.2 Diagnosis of defects and deterioration

Cause	Symptoms			Age of appearance	
	Cracking	Spalling	Erosion	Early	Long-term
Structural deficiency	×	×		×	×
Reinforcement corrosion	×	×			×
Chemical attack	×	×	×		×
Frost damage	×	×	×	×	
Fire damage	×	×		×	
Internal reactions	×	×			×
Thermal effects	×	×		×	×
Shrinkage	×			×	×
Creep	×	×			×
Rapid drying	×			×	
Plastic settlement	×			×	
Physical damage	×	×	×	×	×

summarized in Table 1.2, which is based on the suggestions of Higgins (21). Observation of concrete surface texture and colour variations may be a useful guide to uniformity, and colour change is a widely recognized indicator of the extent of fire damage.

Visual inspection is not confined to the surface, but may also include examination of bearings, expansion joints, drainage channels, post-tensioning ducts and similar features of a structure. Binoculars, telescopes and borescopes may be useful where access is difficult and portable ultra-violet inspection systems may be useful in identifying alkali–aggregate reactions (see section 9.11.1). Recently there has been an increasing acceptance of 'unconventional' methods such as abseiling and robotics to provide cost-effective inspection and remediation access (22). For existing structures, the existence of some feature requiring further investigation is generally initially indicated by visual inspection, and it must be considered the single most important component of routine maintenance. Recent RILEM (6) proposals attempt to provide a numerical classification system to permit the quantification of visual features to assist planning and prioritization. Visual inspection will also provide the basis of judgments relating to access and safety requirements (22) when selecting test methods and test locations.

1.4.3 *Test selection*
Test selection for a particular situation will be based on a combination of factors such as access, damage, cost, speed and reliability, but the basic features of visual inspection followed by a sequence of tests according to convenience and suitability will generally apply. The use of combinations of test methods is discussed in section 1.7.

Testing for durability including causes and extent of deterioration. Relative features of various test methods are summarized in Table 1.3. Corrosion risk of embedded reinforcement is related to the loss of passivity which is provided by the alkaline concrete environment. This is usually as a result of carbonation or chlorides. Simple initial tests will thus involve localized measurements of reinforcement cover, carbonation depths and chloride concentrations. These may be followed by more complex half-cell potential and resistivity testing to provide a more comprehensive survey of large areas. If excessive carbonation is found to be the cause of deterioration, then chemical or petrographic analysis and absorption tests may follow if it is necessary to identify the reasons for this. Direct measurement of the rate of corrosion of reinforcing steel is slowly gaining acceptance as an effective means of assessing the severity of ongoing durability damage and has the potential for use to predict the remaining service lifetime of a corrosion-afflicted structure.

Surface absorption and permeability tests are important in relation to corrosion since both oxygen and water are required to fuel the process, and carbonation rates are also governed by moisture conditions and the ability of atmospheric carbon dioxide to pass through the concrete surface zone. Most other forms of deterioration are also related to moisture which is needed to transport aggressive chemicals and to fuel reactions, thus moisture content, absorption and permeability measurements may again be relevant. Expansion tests on samples of concrete may indicate future performance, and chemical and petrographic testing to assess mix components may be required to identify the causes of disruption of concrete (23).

Table 1.3 Durability tests — relative features

Method	Cost	Speed of test	Damage	Applications
Cover measurement	Low	Fast	None	Corrosion
Carbonation depth	Low	Fast	Minor	risk and
Chloride content	Low	Fast	Minor	cause
Half-cell potential	Moderate/high	Fast	Minor	Corrosion
Resistivity	Moderate/high	Fast	Minor/none	risk
Linear polarization resistance	Moderate/high	Moderate	Minor	Corrosion rate
A.C. impedance	Moderate/high	Slow	Minor	evalution
Galvanostatic pulse	Moderate/high	Fast	Minor	
Absorption	Moderate	Slow	Moderate/minor	Cause and
Permeability	Moderate	Slow	Moderate/minor	risk of
Moisture content	Low	Slow	Minor	corrosion
Chemical	High	Slow	Moderate	and concrete
Petrographic	High	Slow	Moderate	deterioration
Expansion	High	Slow	Moderate	
Radiography	High	Slow	None	

Testing for concrete strength. Relative features of various concrete strength test methods are summarized in Table 1.4.

In the common situation where an assessment of material strength is required, it is unfortunate that the complexity of correlation tends to be greatest for the test methods which cause the least damage. Although surface hardness and pulse velocity tests cause little damage, are cheap and quick, and are ideal for comparative and uniformity assessments, their correlation for absolute strength prediction poses many problems. Core tests provide the most reliable in-situ strength assessment but also cause the most damage and are slow and expensive. They will often be regarded as essential, and their value may be enhanced if they are used to form a basis for calibration of non-destructive or partially-destructive methods which may then be adopted more widely. Whilst most test methods can be successfully applied to concretes made with lightweight aggregates, strength correlations will always be different from those relating to concretes with normal aggregates (24). Partially-destructive methods generally require less-detailed calibration for strength but cause some surface damage, test only the surface zone, and may suffer from high variability. The availability and reliability of strength

Table 1.4 Strength tests — relative merits

Test method	Cost	Speed of test	Damage	Representativeness	Reliability of absolute strength correlations
General applications					
Cores	High	Slow	Moderate	Moderate	Good
Pull-out Penetration resistance	Moderate	Fast	Minor	Near surface only	Moderate
Pull-off Break-off	Moderate	Moderate	Minor	Near surface only	Moderate
Internal fracture	Low	Fast	Minor	Near surface only	Moderate
Comparative assessment					
Ultrasonic pulse velocity	Low	Fast	None	Good	Poor
Surface hardness	Very low	Fast	Unlikely	Surface only	Poor
Strength development monitoring					
Maturity	Moderate		Very minor	Good	Moderate
Temperature-matched curing	High		Very-minor	Good	Good

correlations and the accuracy required from the strength predictions may be important factors in selecting the most appropriate methods to use. This must be coupled with the acceptability of making good any damaged areas for appearance and structural integrity.

When comparison with concrete of similar quality is all that is necessary, the choice of test will be dominated by the practical limitations of the various methods. The least destructive suitable method will be used initially, possibly with back-up tests using another method in critical regions. For example, surface hardness methods may be used for new concrete, or ultrasonics where two opposite surfaces are accessible. When there is only one exposed face, penetration resistance testing is quick and suitable for large members such as slabs, but pull-out or pull-off tests may be more suitable for smaller members. Pull-out testing is particularly useful for direct in-situ measurements of early age strength development, while maturity and temperature-matched curing techniques are based on measurements of within-pour temperatures.

Testing for comparative concrete quality, and localized integrity. Comparative testing is the most reliable application of a number of methods for which calibration to give absolute values of a well-defined physical parameter is not easy. In general, these methods cause little or no surface damage, and most are quick to use, enabling large areas to be surveyed systematically. Some do, however, require relatively complex and expensive equipment.

The most widely used methods are surface hardness, ultrasonic pulse velocity and chain dragging or surface tapping. The latter is particularly useful in locating delamination near to the surface and has been developed with more complex impact-echo techniques. Surface-scanning radar and infrared thermography are both sophisticated methods of locating hidden voids, moisture and similar features which have recently grown in popularity; radiography and radiometry may also be used. Wear tests, surface hardness measurements or surface absorption methods may be use to assess surface abrasion resistance, and thermoluminescence is a specialized technique to assess fire damage.

Testing for structural performance. Large-scale dynamic response testing is available to monitor structural performance, but large-scale static load tests, possibly in conjunction with monitoring of cracking by acoustic emission, may be more appropriate despite the cost and disruption.

Static load tests usually incorporate measurement of deflections and cracking, but problems of isolating individual members can be substantial. Where large numbers of similar elements (such as precast beams) are involved, it may be better to remove a small number of typical elements for laboratory load testing and to use non-destructive methods to compare these elements with those remaining in the structure.

It is essential that the test programme relates the costs of the various test methods to the value of the project involved, the costs of delays to construction, and the cost of possible remedial works. Accessibility of the suspect concrete and the handling of test equipment must be considered, together with the safety of site personnel and the general public during testing operations. Typical examples of test programmes suggested for particular situations are included in Appendix A.

1.4.4 *Number and location of tests*

Establishing the most appropriate number of tests is a compromise between accuracy, effort, cost and damage. Test results will relate only to the specific locations at which the readings or samples were obtained. Engineering judgement is thus required to determine the number and location of tests, and the relevance of the results to the element or member as a whole. The importance of integration of planning with interpretation is thus critical. A full understanding of concrete variability (as discussed in section 1.5) is essential, as well as a knowledge of the reliability of the test method used. This is discussed here with particular reference to concrete strength, since many other properties are strength-related. This should provide a useful general basis for judgments, and further guidance is contained in the chapters dealing with the various test methods. If aspects of durability are involved, care should be taken to allow for variations in environmental exposure and test conditions. Corrosion activity may vary significantly with ambient fluctuations in temperature and rainfall. Care should be taken when estimating mean annual behaviour on the basis of measurements taken on a single occasion. Test positions must also take into account the possible effects of reinforcement upon results, as well as any physical restrictions relating to the method in use.

Table 1.5 lists the number of tests which may be considered equivalent to a single result. The accuracy of strength prediction will depend in most cases on the reliability of the correlation used, but for 'standard' cores 95%

Table 1.5 Relative numbers of readings recommended for various test methods

Test method	No. of individual readings recommended at a location
'Standard' cores	3
Small cores	9
Schmidt hammer	12
Ultrasonic pulse velocity	1
Internal fracture	6
Windsor probe	3
Pull-out	4
Pull-off	6
Break-off	5

confidence limits may be taken as $\pm(12/\sqrt{n})\%$ where n is the number of cores from the particular location. Statistical methods taking acount of the number of tests, test variability and material variability have been developed and are considered more fully in section 1.6.3. Where cores are being used to provide a direct indication of strength or as a basis of calibration for other methods, it is important that sufficient are taken to provide an adequate overall accuracy. It is also essential to remember that the results will relate only to the particular location tested, thus the number of locations to be assessed will be a further factor requiring consideration.

For comparative purposes the truly non-destructive methods are the most efficient, since their speed permits a large number of locations to be easily tested. For a survey of concrete within an individual member, at least 40 locations are suggested, spread on a regular grid over the member, whereas for comparison of similar members a smaller number of points on each member, but at comparable positions, should be examined. Where it is necessary to resort to other methods such as internal fracture or Windsor probe tests, practicalities are more likely to restrict the number of locations examined, and the survey may be less comprehensive.

In-situ strength estimates determining structural adequacy should ideally be obtained for critically stressed locations, in the light of anticipated strength distributions within members (described in section 1.5.1). Attention will thus often be concentrated on the upper zones of members, unless particular regions are suspect.

Tests for material specification compliance must be made on typical concrete, and hence the weaker top zones of members should be avoided. Testing at around mid-height is recommended for beams, columns and walls, and surface zone tests on slabs must be restricted to soffits unless the top layer is first removed. Care must similarly be taken to discard material from the top 20% (or at least 50 mm) of slabs when testing cores.

Where specification compliance is being investigated, it is recommended that no fewer than four cores be taken from the suspect batch of concrete. Where small cores are used, a larger number will be required to give a comparable accuracy, due to greater test variability, and probably at least 12 results are required. With other test methods, a minimum number of readings is less clearly defined but should reflect the values given in Table 1.5 coupled with the calibration reliability. Likely maximum accuracies are summarized in section 1.6. It is inevitable that a considerable 'grey' or 'not proven' area will exist when comparing strength estimates from in-situ testing with specified cube or cylinder strengths, and a best possible accuracy of $\pm 15\%$ has been suggested for a group of four cores (25). This value may increase when dealing with old concrete, due to uncertainties about age effects on strength development. Tests may, however, sometimes be necessary on areas which show signs of poor compaction or workmanship for comparison with other aspects of specifications.

The number of load tests that can be undertaken on a structure will be limited, and these should be concentrated on critical or suspect areas. Visual inspection and non-destructive tests may be valuable in locating such regions. Where individual members are to be tested destructively to provide a calibration for non-destructive methods, they should preferably be selected to cover as wide a range of concrete quality as possible.

1.5 In-situ concrete variability

It is well established that the properties of in-situ concrete will vary within a member, due to differences of compaction and curing as well as non-uniform supply of material. Supply variations will be assumed to be random, but compaction and curing variations follow well-defined patterns according to member type. A detailed appreciation of these variations is essential to planning any in-situ test programme and also to permit sensible interpretation of results.

The average in-situ strength of a member, expressed as the strength of an equivalent cube, will almost invariably be less than that of a standard cube of the same concrete which has been properly compacted and moist-cured for 28 days. The extent of the difference will depend upon materials characteristics, construction techniques, workmanship and exposure, but general patterns can be defined according to member type. This aspect, which is particularly important for interpretation of test results, is discussed in detail in section 1.5.2.

1.5.1 *Within-member variability*
Variations in concrete supply will be due to differences in materials, batching, transport and handling techniques. These will reflect the degree of control over production and will normally be indicated by control and compliance test specimens in which other factors are all standardized. In-situ measurement of these variations is difficult because of the problem of isolating them from compaction and curing effects. They may however be roughly assessed by consideration of the coefficient of variation of tests taken at a number of comparable locations within a member or structure. Compaction and curing effects will depend partially upon construction techniques but are also closely related to member types and location within the member.

Reinforcement may hinder compaction but there will be a tendency for moisture to rise and aggregate to settle during construction. Lower levels of members will further be compacted due to hydrostatic effects, related to member depth, with the result that the general tendency will be for strengths to be highest near the base of pours and lowest in the upper regions. The basic aim of curing is to ensure that sufficient water is present to enable hydration to proceed. For low water: cement ratio mixes, self-desiccation must be avoided by allowing water ingress, and for other mixes, drying out must be prevented. Incomplete hydration resulting from poor curing may cause variations of strength between interior and surface zones of members. A figure of only 5-10% has been suggested for this effect in gravel concretes (26); higher values may apply to lightweight concretes (27). Temperature rises due to cement hydration may cause further strength differences between the interior and outer regions, especially at early ages. Differential curing across members may serve to further increase the variations from compactional factors.

Typical relative strength variations for normal concretes according to member type are illustrated in Figure 1.3. These results have been derived from numerous reports of non-destructive testing including that by Maynard and Davis (28) and can only be regarded as indicating general trends which may be expected, since individual construction circumstances may vary widely. For beams and walls the strength gradients will be reasonably uniform, although variations in compaction and supply may cause the type of variability indicated by the relative strength contours of Figures 1.4 and 1.5. Few data are available for slabs, but it has been suggested that the reduced differential of about 25% across the depths may be concentrated in the top 50 mm in thin slabs (26). Thicker slabs will be more similar to beams. Variations in plan may, however, be expected to be random due to compaction

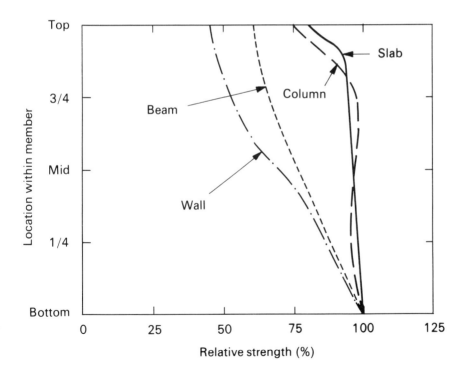

Figure 1.3 Within-member variations.

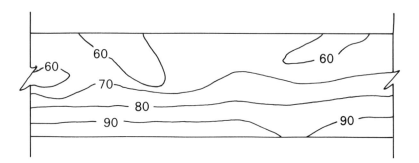

Figure 1.4 Typical relative percentage strength contours for a beam.

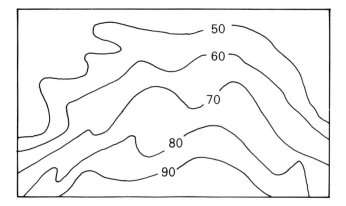

Figure 1.5 Typical relative percentage strength contours for a wall.

and supply inconsistencies. Columns may be expected to be reasonably uniform except for a weaker zone in the top 300 mm or 20% of their depth (29).

It is important to recognize that non-standard concretes may be expected to behave in a manner different from those described above. In particular, Miao *et al.* (30) have demonstrated that high-strength concretes (up to 120 N/mm² cylinder strength) exhibit significantly smaller strength reductions over the height of 1 m-square columns than a 35 N/mm² concrete, which was shown to be reasonably consistent with Figure 1.3. General in-situ variability at a particular height was also found to be smaller at high strengths. Lightweight aggregate concretes have also been shown to have smaller within-depth variations in beams than gravel concrete according to aggregate type and the nature of fine material used (27). This is illustrated in Figure 1.6, which also incorporates differences in in-situ strength relative to 'standard' cube strength (24), discussed in section 1.5.2 below. The most significant reduction in variation can be seen to have occurred when lightweight fines were used, and general within-member variability was also reduced in this case.

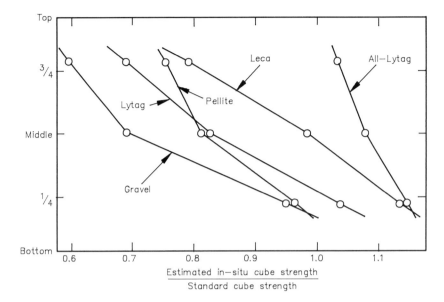

Figure 1.6 Average relative strength distributions within beams of different concrete types (based on ref. 24).

1.5.2 *In-situ strength relative to standard specimens*

Likely strength variations within members have been described in section 1.5.1. If measured in-situ values are expressed as equivalent cube strengths, it will usually be found that they are less than the strengths of cubes made of concrete from the same mix which are compacted and cured in a 'standard' way. In-situ compaction and curing will vary widely, and other factors such as mixing, bleeding and susceptibility to impurities are difficult to predict. Nevertheless a general trend according to member type can be identified and the values given in Table 1.6 may be regarded as typical. Although these are generally accepted (12), cases have been reported where in-situ strengths were found to be closer to that of standard specimens (31) and this is also likely for lightweight aggregate concretes (see Figure 1.6). The likely relationships between standard specimen strength and in-situ strength are also illustrated in Figure 1.7 for a typical structural concrete mix using natural aggregates.

A 'standard' cube is tested whilst saturated, and for ease of comparison the values of Table 1.6 are presented on this basis also. Dry cubes generally yield strengths which are approximately 10–15% higher, and this must be appreciated when interpreting in-situ strength test results. Cores will be tested while saturated under normal circumstances, and the above relationships will apply, but if the in-situ concrete is dry the figures for likely in-situ strength must be increased accordingly. Where non-destructive or partially-destructive

Table 1.6 Comparison of in-situ and 'standard' cube strengths

Member type	Typical 28-day in-situ equivalent wet cube strength as % of 'standard' cube strength	
	Average	*Likely range*
Column	65%	55%–75%
Wall	65%	45%–95%
Beam	75%	60%–100%
Slab	50%	40%–60%

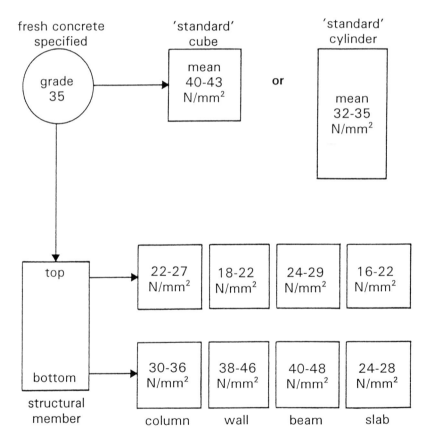

Typical in-situ equivalent 28-day cube strength

Figure 1.7 Typical relationship between standard specimen and in-situ strengths.

methods are used in conjunction with a strength calibration, it is essential to know whether this calibration is based on wet or dry specimens. Another feature of such calibrations is the size of cube upon which they are based. Design and specification are usually based on a 150-mm cube, but laboratory calibrations may sometimes be related to a 100-mm cube which may be up to 4% stronger.

The age at which the concrete is tested is a further cause of differences between in-situ and 'standard' values. Although 'age correction' factors are given in Codes of Practice, care is needed when attempting to adjust in-situ measurements to an equivalent 28-day value. Developments in cement manufacture have tended towards yielding a high early strength with reduced long-term increases, and strength development also is largely dependent on curing. If concrete is naturally wet the strength may increase, but often concrete is dry in service and unlikely to make significant gains after 28 days.

The incorporation of cement replacements such as pulverized fuel ash or ground granulated blast-furnace slag into the mix will also influence longer-term strength development characteristics, and age adjustments should be treated with caution.

1.6 Interpretation

Interpretation of in-situ test results may be considered in three distinct phases leading to the development of conclusions:

(i) Computation
(ii) Examination of variability
(iii) Calibration and/or application.

The emphasis will vary according to circumstances (detailed interpretative information is given in other chapters) but the principles will be similar whatever procedures are used, and these are outlined below. The examples of Appendix A further illustrate the application of those procedures to a number of commonly occurring situations.

The need for comprehensive and detailed recording and reporting of results is of considerable significance, no matter how small or straightforward the investigation may at first appear to be. In the event of subsequent dispute or litigation, the smallest detail may be crucial and documentation should always be kept wth this in mind. Comprehensive photographs are often of particular value for future reference. In-situ test results are also increasingly being incorporated into computer databases, associated with prioritization and management of maintenance and repair strategies (16).

1.6.1 *Computation of test results*
The amount of computation required to provide the appropriate parameter at a test location will vary according to the test method but will follow

well-defined procedures. For example, cores must be corrected for length, orientation and reinforcement to yield an equivalent cube strength.

Pulse velocities must be calculated making due allowance for reinforcement, and pull-out, penetration resistance and surface hardness tests must be averaged to give a mean value. Attempts should not be made at this stage to invoke correlations with a property other than that measured directly. Chemical or similar tests will be evaluated to yield the appropriate parameter such as cement content or mix proportions. Load tests will usually be summarized in the form of load/deflection curves with moments evaluated for critical conditions, and creep and recovery indicated as described in Chapter 6.

1.6.2 *Examination of variability*

Whenever more than one test is carried out, a comparison of the variability of results can provide valuable information. Even where few results are available (e.g. in load tests), these provide an indication of the uniformity of the construction and hence the significance of the results. In cases where more numerous results are available, as in non-destructive surveys, a study of variability can be used to define areas of differing quality. This can be coupled with a knowledge of test variability associated with the method to provide a measure of the construction standards and control used.

Tomsett (32) has also reported the development of an analysis procedure for use on large-scale integrity assessment projects involving a coefficient of variation ratio relating local variability to expected values, an area factor relating the area of the assessed problem to the total area and a comparative damage factor. Interpretation is facilitated by the use of interaction diagrams incorporating these three parameters. Some test methods such as radar and impact-echo rely on recognition of characteristic patterns of test results, and the possibilities for application of neural networks to such cases are currently being studied.

1.6.2.1 Graphical methods. 'Contour' plots showing, for example, zones of equal strength (Figures 1.4 and 1.5) are valuable in locating areas of concrete which are abnormally high or low in strength relative to the remainder of the member. Such contours should be plotted directly on the basis of the parameter measured (e.g. pulse velocity) rather than after conversion to strength. Under normal circumstances the contours will follow well-defined patterns, and any departure from this pattern will indicate an area of concern. 'Contour' plots are also valuable in showing the range of relative strengths within a member and may assist the location of further testing which may be of a more costly or damaging nature. The use of contours is not restricted to strength assessment and they are commonly used for reinforcement corrosion and integrity surveys.

Concrete variability can also be usefully expressed as histograms, especially where a large number of results is available, as when large members are

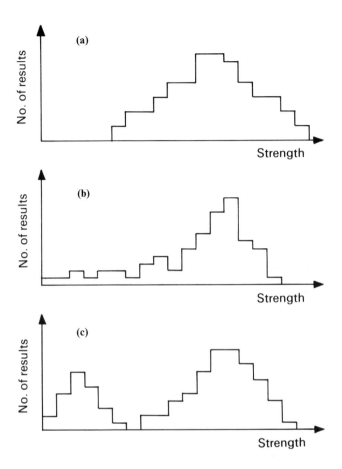

Figure 1.8 Typical histogram plots of in-situ test results: (a) uniform supply; (b) poor construction; (c) two sources.

under test or where many similar members are being compared. Figure 1.8(a) shows a typical plot for well-constructed members using a uniform concrete supply. The parameter measured should be plotted directly, and although the spread will reflect member type and distribution of test locations as well as construction features, a single peak should emerge with an approximately normal distribution. A long 'tail' as in Figure 1.8(b) suggests poor construction procedures, and twin peaks, Figure 1.8(c), indicate two distinct qualities of concrete supply.

1.6.2.2 Numerical methods. Calculation of the coefficient of variation (equal to the standard deviation × 100/mean) of test results may provide valuable information about the construction standards employed. Table 1.7

Table 1.7 Typical coefficients of variation (COV) of test results and maximum accuracies of in-situ strength prediction for principal methods

Test method	Typical COV for individual member of good quality construction	Best 95% confidence limits on strength estimates
Cores — 'standard'	10%	±10% (3 specimens)
'small'	15%	±15% (9 specimens)
Pull-out	8%	±20% (4 tests)
Internal fracture	16%	±28% (6 tests)
Pull-off	8%	±15% (6 tests)
Break-off	9%	±20% (5 tests)
Windsor probe	4%	±20% (3 tests)
Ultrasonic pulse velocity	2.5%	±20% (1 test)
Rebound hammer	4%	±25% (12 tests)

contains typical values of coefficients of variation relating to the principal test methods which may be expected for a single site-made unit constructed from a number of batches. This information is based on the work of Tomsett (33), the authors (26), Concrete Society Report 11 (25) and other sources. Results for concrete from one batch would be expected to be correspondingly lower, whereas if a number of different member types are involved, the values may be expected to be higher. The values in Table 1.7 offer only a very approximate guide, but they should be sufficient to detect the presence of abnormal circumstances.

The coefficient of variation of concrete strength is not constant with varying strength for a given level of control because it is calculated using the average strength. Leshchinsky *et al.* (34) have also confirmed that the distribution of within-test coefficient of variation is asymmetrical. Hence general relationships between coefficient of variation of measured concrete strength and level of construction quality should not be used. Figure 1.9 illustrates typical relationships for 'standard' control cubes and in-situ strengths based on a variety of European and North American sources. From these values, anticipated standard deviations can be deduced (for example at 30 N/mm² mean in-situ strength, a standard deviation of $0.2 \times 30 = 6$ N/mm² is likely for normal quality construction) and hence confidence limits can be placed on the results obtained. Values such as those given later in Table 1.8 can be derived in this way, and in-situ strength accuracy predictions must make allowance for this as well as the accuracy of the test method.

1.6.3 Calibration and application of test results
The likely accuracies of calibration between measured test results and desired concrete properties are discussed in detail in the sections of this book dealing with each specific test. It is essential that the application of the results of in-situ testing takes account of such factors to determine their significance.

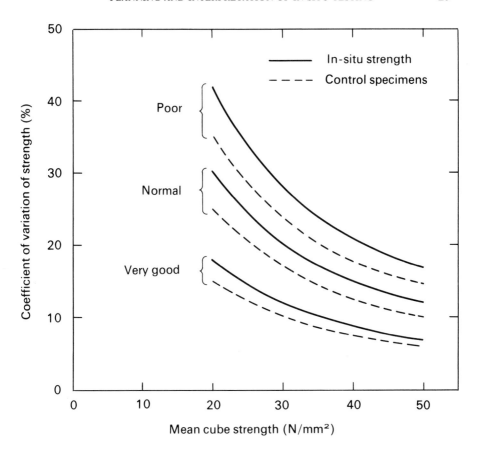

Figure 1.9 Coefficient of variation of test results related to concrete strength.

Particular attention must be paid to the differences between laboratory conditions (for which calibration curves will normally be produced) and site conditions. Differences in maturity and moisture conditions are especially relevant in this respect. Concrete quality will vary throughout members and may not necessarily be identical in composition or condition to laboratory specimens. Also, the tests may not be so easy to perform or control due to adverse weather conditions, difficulties of access or lack of experience of operatives. Calibration of non-destructive and partially-destructive strength tests by means of cores from the in-situ concrete may often be possible and will reduce some of these differences.

Interpretation of strength results requires the use of statistical procedures since it is not sufficient simply to average the values of the in-situ test results and then compute the equivalent compressive strength by means of the previously established relationship. Efforts have been made to establish lower confidence limits for the correlation relationship (1, 35) based on statistical tolerance factors, and the procedures outlined in the following sections are based on this relatively simple approach. These methods fail however to take account of measurement errors in the in-situ test result, as demonstrated by Stone *et al.* (36). A rigorous method incorporated into a report by ACI 228 (37) in 1989 has not been used extensively because of its complexity, but a simplified version (38) is likely to be incorporated into a forthcoming revision.

The current lack of a consensus-based statistical procedure is a barrier to more widespread use of in-place testing for compliance purposes. Leshchinsky (39) has reviewed current provisions of existing national standards and the matter is under consideration by RILEM Committee 126 at present.

Table 1.7 summarizes the maximum accuracies of in-situ strength prediction that can realistically be hoped for under ideal conditions, with specific calibrations for the particular concrete mix in each case. If any factor varies from this ideal, the accuracies of prediction will be reduced, although at present there is little available information to permit this to be quantified. Wherever possible, test methods should be used which directly measure the required property, thereby reducing the uncertainties involved in calibration. Even in these situations, however, care must be taken to make a realistic assessment of the accuracy of the values emerging when formulating conclusions.

1.6.3.1 Application to specifications.   It is essential that the concrete tested is representative of the material under examination and this will influence the number and location of tests (section 1.4.4). Where some clearly defined property, such as cover or cement content, is being measured, it will generally be sufficient to compare measured results with the minimum specified value, bearing in mind the likely accuracy of the test. A small proportion of results marginally below the specified value may be acceptable, but the average for a number of locations should exceed the minimum limit. If the test has a low order of accuracy (for example cement content determination is unlikely to be better than $\pm 40 \, \text{kg/m}^3$) the area of doubt concerning marginal results may be considerable. This is an unfortunate fact of life, although engineering judgements may perhaps be assisted by corroborative measurements of a different property.

Strength is the most common criterion for the judgement of compliance with specifications, and unfortunately the most difficult to resolve from in-situ testing because of the basic differences between in-situ concrete and the 'standard' test specimens upon which most specifications are based (section 1.5.2). The number of in-situ test results will seldom be sufficient to permit a full statistical assessment of the appropriate confidence limits (usually 95%), hence it is better to compare mean in-situ strength estimates with the expected mean 'standard' test specimen result. This requires an estimate to be made of the likely standard deviation of standard specimens unless the value of target mean strength for the mix is known. The mean 'standard' cube strength using British 'limit state' design procedures is given by

$$f_{mean} = f_{cu} + 1.64s \qquad (1.1)$$

where f_{cu} = characteristic strength of control cubes
$\qquad s$ = standard deviation of control cubes.

The accuracy of this calculation will increase with the number of results available; 50 readings could be regarded as the minimum necessary to obtain a sufficiently accurate estimate of the actual standard deviation. If sufficient information is not available the values given in Table 1.8 may be used as a guide.

In theory it is possible to estimate the in-situ characteristic strength f'_{cu} from the measured in-situ values of the mean f'_{mean} and standard deviation s'. The values of s' given in Table 1.8 may be used in the absence of more specific data, but cannot be considered very reliable in view of within-member variations and the many variable constructional factors.

In most cases the number of readings available from in-situ results will be significantly less than 50, in which case the coefficient of 1.64 used in eqn (1.1) will increase. Equation (1.2) for the 95% confidence limit will thus apply, with k given by Table 1.9 according to the number of results n.

$$f'_{cu} = f'_{mean} - ks' \qquad (1.2)$$

Table 1.8 Typical values of standard deviation of control cubes and in-situ concrete

Material control and construction	Assumed std. devn. of control cube (s) (N/mm^2)	Estimated std. devn. of in-situ concrete (s') (N/mm^2)
Very good	3.0	3.5
Normal	5.0	6.0
Low	7.0	8.5

Table 1.9 Suggested 95% confidence limit factor related to number of tests (13)

Number of tests n	Confidence factor k
3	10.31
4	4.00
5	3.00
6	2.57
8	2.23
10	2.07
12	1.98
15	1.90
20	1.82
∞	1.64

This equation assumes a 'normal' distribution of concrete strength results (as in Eqn (1.1)) but where concrete variability is high, as for poor quality control, a 'log-normal' distribution is considered to be more realistic (37). In this case

$$\log f'_{cu} = \text{mean value of } [\log f'] - k \times \text{standard deviation of } [\log f'] \quad (1.3)$$

where f' is an individual in-situ strength result.

These relationships can conveniently be represented in graphical form as in Figure 1.10, which can be used to evaluate the characteristic value as a proportion of the mean for a particular coefficient of variation of results. In this figure 'normal' and 'log-normal' distributions are compared directly for a coefficient of variation of 15% and the less demanding nature of the 'log-normal' distribution is demonstrated. This effect increases with increasing coefficient of variation. The combined effects of variability of results and number of tests can also be clearly seen and the importance of having at least four results is apparent. Bartlett and MacGregor have applied this approach to the evaluation of equivalent in-place characteristic strength from core test data (40).

Where some indications of the expected mean and material variability are available, a preliminary calculation can be made to obtain the desired characteristic strength as a proportion of the mean, and hence the minimum

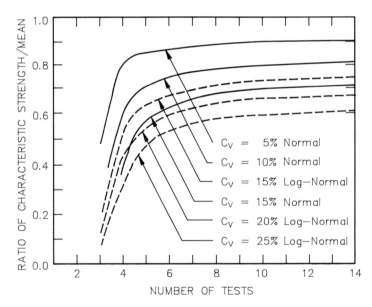

Figure 1.10 Characteristic strength (95% confidence limit) as a function of coefficient of variation and number of tests (based on ref. 13).

number of tests required to confirm the desired acceptability can be evaluated
(13). Similar plots can be produced for different confidence limits and
distributions (37) and it should be noted that less demanding 90% confidence
limits are adopted in some countries. The choice of distribution type and
confidence limits for use in particular circumstances is thus a matter of judgement.

If an in-situ characteristic strength is estimated it can be compared with
the specified value, but this approach is not recommended unless numerous
in-situ results are available.

Whichever approach is adopted, the comparison between in-situ and
standard specimen strengths must allow for the type of differences indicated
in Table 1.6 and Figure 1.7 and this is illustrated in the examples of Appendix A.

1.6.3.2 Application to design calculations. Measured in-situ values can be
incorporated into calculations to assess structural adequacy. Although this
may occasionally relate to reinforcement quantities and location, or concrete
properties such as permeability, in most instances it will be the concrete
strength which is relevant. It is essential that the measured values relate to
critical regions of the member under examination and tests must be planned
with this in mind (section 1.4.4).

Calculations are generally based on minimum likely, or characteristic,
'standard specimen' values being modified by an appropriate factor of safety
to give a minimum in-situ design value. In-situ measurements will yield
directly an in-situ strength of the concrete tested and this must be related to
a similar specimen type and size to the 'standard' used in the calculations. If
this concrete is from a critical location, it could be argued that the minimum
measured value can be used directly as the design concrete strength with no
further factors of safety applied. In practice, however, it is more appropriate
to use the mean value from a number of test readings at critical locations,
and to apply a factor of safety to this to account for test variability, possible
lack of concrete homogeneity and future deterioration. The accuracy of
strength prediction will vary according to the method used, but a factor of
safety of 1.2 is recommended by BS 6089 (12) for general use. Providing the
recommendations of section 1.4.4 have been followed when determining the
number of readings, this value should be adequate. The application of this
approach is illustrated in detail by the examples of Appendix A. If there is
particular doubt about the reliability of the test results, or if the concrete
tested is not from the critical location considered, then it may be necessary
for the engineer to adopt a higher value for the factor of safety guided by
the information contained in sections 1.5.1, 1.5.2 and 1.6.3.1. Alternatively,
other features discussed in section 1.5.2, including moisture condition and
age, may possibly be used to justify a lower value for the factor of safety.
The in-situ stress state and rate of loading may also be taken into account
in critical circumstances.

1.7   Test combinations

All the test methods which are available for in-situ concrete assessment suffer from limitations, and reliability is often open to question. Combining methods may help to overcome some of these difficulties, and some examples of typical combinations are outlined below.

1.7.1   *Increasing confidence level of results*

Considerably greater weight can be placed on results if corroborative conclusions can be obtained from separate methods. Expense will usually restrict large-scale duplication, but if different properties are measured, confidence will be much increased by the emergence of similar patterns of results. This will generally be restricted to tests which are quick, cheap and non-destructive, such as combinations of surface hardness and ultrasonic pulse velocity measurements on recently cast concrete. In other circumstances, radiometry, pulse-echo, radar, thermography, or slower near-to-surface strength methods may be invaluable.

If small volumes are involved and a specific property (e.g. strength) is required, it may sometimes be worth while to compare absolute estimates achieved by different methods.

1.7.2   *Improvement of calibration accuracy*

It may, in some cases, be possible to produce correlations of combinations of measured values with desired properties, to a greater accuracy than is possible for either individual method. This has been most widely developed in relation to strength assessment using ultrasonic pulse velocities in conjunction with density (41) or rebound hammer readings (which are related to surface density).

In the latter case, appropriate strength correlations must be produced for both methods enabling multiple regression equations to be developed with compressive strength as the dependent variable (42). This approach is likely to be of greatest value in quality control situations but is not widely used. A more complex version of the technique has been encompassed in the SONREB method as a draft RILEM recommendation (43). This is based largely on work in eastern Europe and involves the principle that correlation graphs may be produced involving coefficients relating to various properties of the mix constituents. The increased accuracy is attributed to the opposing influences of some of the many variables for each of the methods, and strength predictions to an accuracy of $\pm 10\%$ are claimed under ideal conditions.

Other combinations that have been proposed include the use of pulse velocity and pulse attenuation measurements on site (44). The procedures are complex and require specialized equipment, and for practical purposes this approach must still be considered as a research tool. The more common in-situ tests may certainly be combined in a variety of other ways but although

Table 1.10 Relevant standards

British standards
BS 1881: Testing concrete

Part 5	Methods of testing concrete for other than strength
Part 120	Determination of compressive strength of concrete cores
Part 124	Chemical analysis of hardened concrete
Part 201	Guide to the use of NDT for hardened concrete
Part 202	Surface hardness testing by rebound hammer
Part 203	Measurement of the velocity of ultrasonic pulses in concrete
Part 204	The use of electromagnetic covermeters
Part 205	Radiography of concrete
Part 206	Determination of strain in concrete
Part 207	Near to surface test methods for strength
*Part 208	Initial surface absorption test
*In preparation	

BS 812

Part 1	Sampling and testing mineral aggregates, sands and filters

BS 6089: Assessment of concrete strength in existing structures
BS 8110: Structural use of concrete
BS DD92: Temperature matched curing of concrete specimens

American standards
ASTM

C42	Standard method of obtaining and testing drilled cores and sawed beams of concrete
C85	Cement content of hardened Portland cement concrete
C457	Air void content in hardened concrete
C597	Standard test method for pulse velocity through concrete
C779	Abrasion resistance of horizontal concrete surfaces
C803	Penetration resistance of hardened concrete
C805	Rebound number of hardened concrete
C823	Examining and sampling of hardened concrete in constructions
C856	Petrographic examination of hardened concrete
C876	Half-cell potential of uncoated reinforcing steel in concrete
C900	Pull-out strength of hardened concrete
C918	Measurement of early-age compressive strength and projecting later age strength
C944	Abrasion resistance of concrete or mortar surfaces by the rotating cutter method
C1040	Density of unhardened and hardened concrete in place by nuclear methods
C1074	Estimating concrete strength by the maturity method
C1150	Break-off number of concrete
D4580	Measuring delaminations in concrete bridge decks by sounding
D4748	Determining the thickness of bound pavement layers using short-pulse radar
D4788	Detecting delaminations in bridge decks using infrared thermography

valuable corroborative evidence may be gained it is unlikely that the accuracy of absolute strength predictions will be significantly improved.

1.7.3 *Use of one method as preliminary to another*
Combinations of methods are widely used in situations where one method is regarded as a preliminary to the other. Common examples include the

location of reinforcement prior to other forms of testing, and the use of simple non-destructive methods for comparative surveys to assist the most worthwhile location of more expensive or damaging tests (see Figure 1.1). Tomsett has reported the successful combination of thermography and ultrasonic pulse velocity measurements used in this way (33).

Where monitoring strength development is important, maturity measurements may provide useful preliminary information, for confirmation by other strength assessment methods. A further case is the use of half-cell potential measurements to indicate the level of possibility of corrosion occurring, and subsequent resistivity measurements on zones shown to be at risk will identify the likelihood of corrosion actually occurring.

1.7.4   Test calibration
The most frequently occurring examples of calibration involving test combinations will be the use of cores or destructive load tests to establish correlations for non-destructive or partially-destructive methods which relate directly to the concrete under investigation. Coring or drilling may also be required to calibrate or validate the results of radar surveys.

1.7.5   Diagnosis of causes of deterioration
It is most likely that more than one type of testing will be required to identify the nature and cause of deterioration, and to assess future durability. Cover measurements will be included if reinforcement corrosion is involved, together with a possible range of chemical, petrographic and absorption tests. Where deterioration is due to disruption of the concrete, a variety of tests on samples removed from the concrete is likely to be required, as discussed in section 1.4.3.

1.8   Documentation by standards

Many British and American Standards now available are applicable to in-situ concrete testing. A selection of those which are most relevant is listed in Table 1.10, and is fully referenced in the appropriate parts of the text elsewhere in this book.

2 Surface hardness methods

One of many factors connected with the quality of concrete is its hardness. Efforts to measure the surface hardness of a mass of concrete were first recorded in the 1930s; tests were based on impacting the concrete surface with a specified mass activated by a standard amount of energy. Early methods involved measurements of the size of indentation caused by a steel ball either fixed to a pendulum or spring hammer, or fired from a standardized testing pistol. Later, however, the height of rebound of the mass from the surface was measured. Although it is difficult to justify a theoretical relationship between the measured values from any of these methods and the strength of a concrete, their value lies in the ability to establish empirical relationships between test results and quality of the surface layer. Unfortunately these are subject to many specific restrictions including concrete and member details, as well as equipment reliability and operator technique.

Indentation testing has received attention in Germany and in former states of the USSR as well as the United Kingdom, but has never become very popular. Pin penetration tests have, however, recently received attention in the USA and Japan (see section 4.1.2). The rebound principle, on the other hand, is more widely accepted: the most popular equipment, the Schmidt Rebound Hammer, has been in use worldwide for many years. Recommendations for the use of the rebound method are given in BS 1881: Part 202 (45) and ASTM C805 (46).

2.1 Rebound test equipment and operation

The Swiss engineer Ernst Schmidt, first developed a practicable rebound test hammer in the late 1940s, amd modern versions are based on this. Figure 2.1 shows the basic features of a typical type N hammer, which weighs less than 2 kg, and has an impact energy of approximately 2.2 Nm.

The spring-controlled hammer mass slides on a plunger within a tubular housing. The plunger retracts against a spring when pressed against the concrete surface and this spring is automatically released when fully tensioned, causing the hammer mass to impact against the concrete through the plunger. When the spring-controlled mass rebounds, it takes with it a rider which slides along a scale and is visible through a small window in the side of the casing. The rider can be held in position on the scale by depressing the locking button. The equipment is very simple to use (Figure 2.2), and may be operated either horizontally or vertically either upwards or downwards.

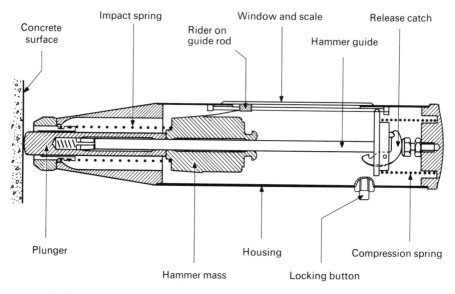

Figure 2.1 Typical rebound hammer.

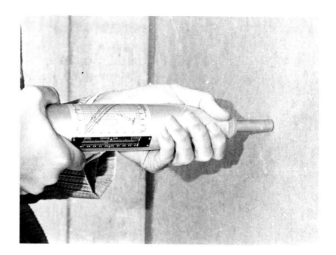

Figure 2.2 Schmidt hammer in use.

The plunger is pressed strongly and steadily against the concrete at right angles to its surface, until the spring-loaded mass is triggered from its locked position. After the impact, the scale index is read while the hammer is still in the test position. Alternatively, the locking button may be pressed to enable

Figure 2.3 Digi-Schmidt (photograph by courtesy of Steinweg UK Ltd).

the reading to be retained, or results can be recorded automatically by an attached paper recorder. The scale reading is known as the rebound number, and is an arbitrary measure since it depends on the energy stored in the given spring and on the mass used. This version of the equipment is most commonly used, and is most suitable for concretes in the $20-60 \, \text{N/mm}^2$ strength range. Electronic digital reading equipment with automatic data storage and processing facilities is also available (Figure 2.3). Other specialized versions are available for impact sensitive zones and for mass concrete. For low strength concrete in the $5-25 \, \text{N/mm}^2$ strength range it is recommended that a pendulum type rebound hammer as shown in Figure 2.4 is used which has an enlarged hammer head (Type P).

2.2 Procedure

The reading is very sensitive to local variations in the concrete, especially to aggregate particles near to the surface. It is therefore necessary to take several readings at each test location, and to find their average. Standards vary in their precise requirements, but BS 1881: Part 202 (45) recommends 12 readings taken over an area not exceeding 300 mm square, with the impact points no less than 20 mm from each other or from an edge. The use of a grid to locate these points reduces operator bias. The surface must be smooth, clean and

Figure 2.4 Pendulum hammer.

dry, and should preferably be formed, but if trowelled surfaces are unavoidable they should be rubbed smooth with the carborundum stone usually provided with the equipment. Loose material can be ground off, but areas which are rough from poor compaction, grout loss, spalling or tooling must be avoided since the results will be unreliable.

2.3 Theory, calibration and interpretation

The test is based on the principle that the rebound of an elastic mass depends on the hardness of the surface upon which it impinges, and in this case will provide information about a surface layer of the concrete defined as no more than 30 mm deep. The results give a measure of the relative hardness of this zone, and this cannot be directly related to any other property of the concrete. Energy is lost on impact due to localized crushing of the concrete and internal friction within the body of the concrete, and it is the latter, which is a function of the elastic properties of the concrete constituents, that makes theoretical evaluation of test results extremely difficult (47). Many factors influence results but must all be considered if rebound number is to be empirically related to strength.

2.3.1 *Factors influencing test results*
Results are significantly influenced by all the following factors:

1. Mix characteristics
    (i) Cement type
    (ii) Cement content
    (iii) Coarse aggregate type

2. Member characteristics
    (i) Mass
    (ii) Compaction
    (iii) Surface type
    (iv) Age, rate of hardening and curing type
    (v) Surface carbonation
    (vi) Moisture condition
    (vii) Stress state and temperature.

Since each of these factors may affect the readings obtained, any attempts to compare or estimate concrete strength will be valid only if they are all standardized for the concrete under test and for the calibration specimens. These influences have different magnitudes. Hammer orientation will also influence measured values (section 2.3.2) although correction factors can be used to allow for this effect.

2.3.1.1 Mix characteristics. The three mix characteristics listed above are now examined in more detail.

   (i) *Cement type.* Variations in fineness of Portland cement are unlikely to be significant — their influence on strength correlation is less than 10%. Super-sulphated cement, however, can be expected to yield strengths 50% lower than suggested by a Portland cement calibration, whereas high alumina cement concrete may be up to 100% stronger.
   (ii) *Cement content.* Changes in cement content do not result in corresponding changes in surface hardness. The combined influence of strength, workability and aggregate/cement proportions leads to a reduction of hardness relative to strength as the cement content increases (48). The error in estimated strength, however, is unlikely to exceed 10% from this cause for most mixes.
   (iii) *Coarse aggregate.* The influence of aggregate type and proportions can be considerable, since strength is governed by both paste and aggregate characteristics. The rebound number will be influenced more by the hardened paste. For example, crushed limestone may yield a rebound number significantly lower than for a gravel concrete of similar strength which may typically be equivalent to a strength difference of 6–7 N/mm². A particular aggregate type may also yield different rebound number/strength correlations depending on the source and nature, and Figure 2.5 compares typical curves for high-

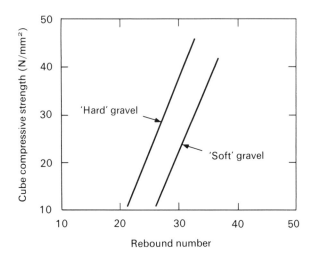

Figure 2.5 Comparison of hard and soft gravels — vertical hammer.

and low-quality gravels. These have measured hardness expressed in terms of the Mohs' number (see section 4.1.1.2) of 7 and 3 respectively.

Lightweight aggregates may be expected to yield results significantly different from those for concrete made with dense aggregates, and considerable variations have also been found between types of lightweight aggregates (24). Calibrations can, however, be obtained for specific lightweight aggregates, although the amount of natural sand used will affect results.

The extent of these differences is illustrated by Figure 2.6 which compares strength correlations obtained by varying age of otherwise 'identical' dry-cured laboratory specimens containing different lightweight coarse aggregates. Mix 5 included lightweight fine materials whilst all others contained natural sand, and the effects of this can be seen by comparing results for mixes 4 and 5 which are otherwise similar.

2.3.1.2 Member characteristics. The member characteristics listed above are also to be discussed in detail.

  (i) *Mass.* The effective mass of the concrete specimen or member under test must be sufficiently large to prevent vibration or movement caused by the hammer impact. Any such movement will result in a reduced rebound number. For some structural members the slenderness or mass may be such that this criterion is not fully satisfied, and in such cases absolute strength prediction may be difficult. Strength comparisons between or within individual members must also take account of this factor. The mass of calibration specimens may be effectively increased

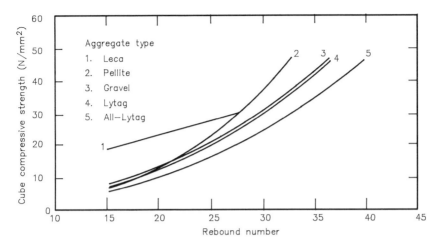

Figure 2.6 Comparison of lightweight aggregates (based on ref. 24).

by clamping them firmly in a heavy testing machine, and this is discussed more fully in section 2.3.2.

(ii) *Compaction.* Since a smooth, well-compacted surface is required for the test, variations of strength due to internal compaction differences cannot be detected with any reliability. All calibrations must assume full compaction.

(iii) *Surface type.* Hardness methods are not suitable for open-textured or exposed aggregate surfaces. Trowelled or floated surfaces may be harder than moulded surfaces, and will certainly be more irregular. Although they may be smoothed by grinding, this is laborious and it is best to avoid trowelled surfaces in view of the likely overestimation of strength from hardness readings. The absorption and smoothness of the mould surface will also have a considerable effect. Calibration specimens will normally be cast in steel moulds which are smooth and non-absorbent, but more absorbent shuttering may well produce a harder surface, and hence internal strength may be overestimated. Although moulded surfaces are preferred for on-site testing, care must be taken to ensure that strength calibrations are based on similar surfaces, since considerable errors can result from this cause.

(iv) *Age, rate of hardening and curing type.* The relationship between hardness and strength has been shown to vary as a function of time (48), and variations in initial rate of hardening, subsequent curing, and exposure conditions will further influence this relationship. Where heat treatment or some other form of accelerated curing has been used, a specific calibration will be necessary. The moisture state may also be influenced by the method of curing. For practical purposes the influence of time may be regarded as unimportant up to the age of three months, but for older concretes it may be possible to develop reduction factors which take account of the concrete's history.

(v) *Surface carbonation.* Concrete exposed to the atmosphere will normally form a hard carbonated skin, whose thickness will depend upon the exposure conditions and age. It may exceed 20 mm for old concrete although it is unlikely to be significant at ages of less than three months. The depth of carbonation can easily be determined as described in Chapter 9. Examination of gravel concrete specimens which had been exposed to an outdoor 'city-centre' atmosphere for six months showed a carbonated depth of only 4 mm. This was not sufficient to influence the rebound number/strength relationship in comparison with similar specimens stored in a laboratory atmosphere, although for these specimens no measurable skin was detected. In extreme cases, however, it is known that the overestimate of strength from this cause may be up to 50%, and is thus of great importance. When significant carbonation is known to exist, the surface layer ceases to be representative of the concrete within an element.

(vi) *Moisture condition.* The hardness of a concrete surface is lower when wet than when dry, and the rebound/strength relationship will be influenced accordingly. This effect is illustrated by Figure 2.7, based on work by the US army (49), from which it will be seen that a wet

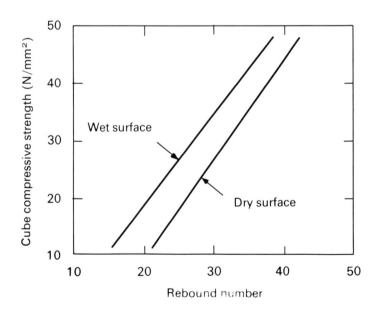

Figure 2.7 Influence of surface moisture condition — horizontal hammer (based on ref. 49).

surface test may lead to an underestimate of strength of up to 20%. Field tests and strength calibrations should normally be based on dry surface conditions, but the effect of internal moisture on the strength of control specimens must not be overlooked. This is considered in more detail in section 2.3.2.

(vii) *Stress state and temperature.* Both these factors may influence hardness readings, although in normal practical situations this is likely to be small in comparison with the many other variables. Particular attention should, however, be paid to the functioning of the test hammer if it is to be used under extremes of temperature.

2.3.2 *Calibration*

Clearly, the influences of the variables described above are so great that it is very unlikely that a general calibration curve relating rebound number to strength, as provided by the equipment manufacturers, will be of any practical value. The same applies to the use of computer data processing to give strength predictions based on results from the electronic rebound hammer shown in Figure 2.3 unless the conversions are based on case-specific data. Strength calibration must be based on the particular mix under investigation, and the mould surface, curing and age of laboratory specimens should correspond as closely as possible to the in-place concrete. It is essential that correct functioning of the rebound hammer is checked regularly using a standard steel anvil of known mass. This is necessary because wear may change the spring and internal friction characteristics of the equipment. Calibrations prepared for one hammer will also not necessarily apply to another. It is probable that very few rebound hammers used for in-situ testing are in fact regularly checked against a standard anvil, and the reliability of results may suffer as a consequence.

The importance of specimen mass has been discussed above; it is essential that test specimens are either securely clamped in a heavy testing machine or supported upon an even solid floor. Cubes or cylinders of at least 150 mm should be used, and a minimum restraining load of 15% of the specimen strength has been suggested for cylinders (50), and BS 1881 (45) recommends not less than 7 N/mm^2 for cubes tested with a type N hammer. Some typical relationships between rebound number and restraining load are given in Figure 2.8, which shows that once a sufficient load has been reached the rebound number remains reasonably constant.

It is well established that the crushing strength of a cube tested wet is likely to be about 10% lower than the strength of a corresponding cube tested dry. Since rebound measurements should be taken on a dry surface, it is recommended that wet cured cubes be dried in the laboratory atmosphere for 24 hours before test, and it is therefore to be expected that they will yield higher strengths than if tested wet in the standard manner (51). Depending

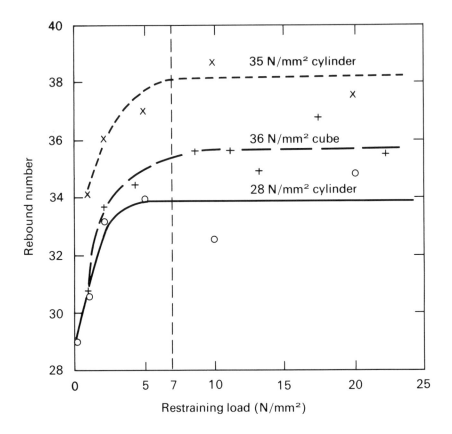

Figure 2.8 Effect of restraining load on calibration specimen (incorporating data from ref. 50).

upon the purpose of the test programme it may be necessary to confirm this relationship, and the relative moisture conditions of the calibration specimens and in-place concrete must also be considered when interpreting the field results. The use of cores cut following in-situ hardness tests may help to overcome these difficulties in developing calibrations.

If cubes are used, readings should be taken on at least two vertical faces of the specimen as cast, as described in section 2.2, and the hammer orientation must be similar to that to be used for the in-place tests. The influence of gravity on the mass will depend on whether it is moving vertically up or down, horizontally or on an inclined plane. The effect on the rebound number will be considerable, although the relative values suggested by the manufacturer are likely to be reliable in this instance because this is purely a function of the equipment.

2.3.3 *Interpretation*

The interpretation of surface hardness readings relies upon a knowledge of the extent to which the factors described in section 2.3.1 have been standardized between readings being compared. This applies whether the results are being used to assess relative quality or to estimate strength. It will be apparent from Figure 2.9, which shows a typical strength calibration chart produced under 'ideal' laboratory conditions, that the scatter of results is considerable, and the strength range corresponding to a given rebound number is about $\pm 15\%$ even for 'identical' concrete. In a practical situation it is very unlikely that a strength prediction can be made to an accuracy better than $\pm 25\%$ (50). The calibration scatter also suggests that even if a strength prediction is not required, a considerable variation of rebound number can be expected for 'identical' concrete, and acceptable limits must be determined in conjunction with some other form of testing. It is suggested (12) that where the total number of readings (n) taken at a location is not less than 10, the accuracy of the mean rebound number is likely to be within $\pm 15/\sqrt{n}\%$ with 95% confidence. The results may usefully be presented in graphical form as described in section 1.6.2.1, and calculation of the coefficient of variation may yield an indication of concrete uniformity, as described in section 1.6.2.2, when sufficient results are available.

The test location within the member is important when interpreting results

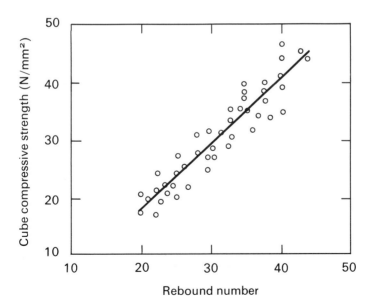

Figure 2.9 Typical rebound number/compressive strength calibration chart.

(Chapter 1) but it should be noted that the test yields information about a thin surface layer only. Results are unrelated to the properties of the interior, and furthermore are not regarded as reliable on concrete more than three months old unless special steps are taken to allow for age effects and surface carbonation, as described above.

Although it is generally the relationship between rebound number and compressive strength that is of interest, similar relationships can be established with flexural strength although with an even greater scatter. It appears that no general relationship between rebound number and elastic modulus exists although it may be possible to produce such a calibration for a specific mix.

2.3.4 *Applications and limitations*

The useful applications of surface hardness measurements can be divided into four categories:

(i) Checking the uniformity of concrete quality
(ii) Comparing a given concrete with a specified requirement
(iii) Approximate estimation of strength
(iv) Abrasion resistance classification.

Whatever the application, it is essential that the factors influencing test results are standarized or allowed for, and it should be remembered that results relate only to the surface zone of the concrete under test. A further overriding limitation relates to testing at early ages or low strengths, because the rebound numbers may be too low for accurate reading and the impact may also cause damage to the surface (Figure 2.10). It is therefore not recommended that the method is used for concrete which has a cube strength of less than $10 \, N/mm^2$ or which is less than 7 days old, unless of high strength.

(i) *Concrete uniformity checking*. The most important and reliable applications of surface hardness testing are where it is not necessary to attempt to convert the results to some other property of the concrete. It is claimed (48) that surface hardness measurements give more consistently reproducible results than any other method of testing concrete. Although they do not detect poor internal compaction, results are sensitive to variations of quality between batches, or due to inadequate mixing or segregation. The value as a control test is further enhanced by the ability to monitor the concrete in members cheaply and more comprehensively than is possible by a small number of control specimens. For such comparisons to be valid for a given mix it is only necessary to standardize age, maturity, surface moisture conditions (which should preferably be dry), and location on the structure or unit.

This approach has been extensively used to control uniformity of precast concrete units, and may also prove valuable for the comparison of suspect in-situ elements with similar elements which are known to be sound. A further valuable use for such comparative tests may be to establish the representation

Figure 2.10 Surface damage on green concrete.

of other forms of testing, possibly destructive, which may yield more specific but localized indications of quality.

(ii) *Comparison with a specific requirement.*   This application is also popular in the precasting industry, where a minimum hardness reading may be calibrated against some specific requirement of the concrete. For instance, the readiness of precast units for transport may be checked, with calibration based on proof load tests. The approach may also be used as an acceptance criterion, in relation to the removal of temporary supports from structural members, or commencement of stress transfer in prestressed concrete construction.

(iii) *Approximate strength estimation.*   This represents the least reliable application and (unfortunately, since a strength estimate is frequently required by engineers) is where misuse is most common. The accuracy depends entirely upon the elimination of influences which are not taken into account in the calibration. For laboratory specimens cast, cured and tested under conditions identical to those used for calibration, it is unlikely that a strength estimate better than $\pm 15\%$ can be achieved for concrete up to three months old. Although it may be possible to correct for one or two variables which may not be identical on site, the accuracy of absolute strength prediction will decline as a consequence and is unlikely to be better than $\pm 25\%$. The use of the rebound hammer for strength estimation of in-place concrete must never be attempted unless specific calibration charts are available, and even then, the use of this method alone is not recommended, although the value of results may be improved if used in conjunction with other forms of testing as described in Chapter 1.

(iv) *Abrasion resistance classification.* Abrasion resistance is generally affected by the same influences as surface hardness, and Chaplin (52) has suggested that the rebound hammer may be used to classify this property. This is discussed in Chapter 7. It is also reasonable to suppose that other durability characteristics that are related to a dense, well cured, outer surface zone may similarly be classified.

3 Ultrasonic pulse velocity methods

The first reports of the measurement of the velocity of mechanically generated pulses through concrete appeared in the USA in the mid 1940s. It was found that the velocity depended primarily upon the elastic properties of the material and was almost independent of geometry. The potential value of this approach was apparent, but measurement problems were considerable, and led to the development in France, a few years later, of repetitive mechanical pulse equipment. At about the same time work was undertaken in Canada and the United Kingdom using electro-acoustic transducers, which were found to offer greater control on the type and frequency of pulses generated. This form of testing has been developed into the modern ultrasonic method, employing pulses in the frequency range of 20–150 kHz, generated and recorded by electronic circuits. Ultrasonic testing of metals commonly uses a reflective pulse technique with much higher frequencies, but this cannot readily be applied to concrete because of the high scattering which occurs at matrix/aggregate interfaces and microcracks. Concrete testing is thus at present based largely on pulse velocity measurements using through-transmission techniques. The method has become widely accepted around the world, and commercially produced robust lightweight equipment suitable for site as well as laboratory use is readily available.

Andrews (53) has suggested that there is much scope for new applications with the development of improved fidelity transducers and computer interpretation. Study of pulse attenuation characteristics has been shown by the authors to provide useful data relating to deterioration of concrete due to alkali–silica reactions (54) although there are practical problems of achieving consistent coupling on site. Hillger (55) and Kroggel (56) have both described the development of pulse-echo techniques to permit detection of defects and cracks from tests on one surface as well as the use of a vacuum coupling system, and the application of signal processing techniques to yield information about internal defects and features is the subject of current research. Another interesting development described by Sack and Olson (57) involves the use of rolling transmitter and receiver scanners which do not need any coupling medium with a computer data acquisition system which permits straight line scans of up to 9 m to be made within a time-scale of less than 30 seconds.

Although it is likely that such developments will expand into commercial use in the near future, the remainder of this chapter will concentrate upon conventional pulse velocity techniques.

If the method is properly used by an experienced operator, a considerable amount of information about the interior of a concrete member can be obtained. However, since the range of pulse velocities relating to practical concrete qualities is relatively small (3.5–4.8 km/s), great care is necessary, especially for site usage. Furthermore, since it is the elastic properties of the concrete which affect pulse velocity, it is often necessary to consider in detail the relationship between elastic modulus and strength when interpreting results. Recommendations for the use of this method are given in BS 1881: Part 203 (58) and also in ASTM C597 (59).

3.1   Theory of pulse propagation through concrete

Three types of waves are generated by an impulse applied to a solid mass. Surface waves having an elliptical particle displacement are the slowest, whereas shear or transverse waves with particle displacement at right angles to the direction of travel are faster. Longitudinal waves with particle displacement in the direction of travel (sometimes known as compression waves) are the most important since these are the fastest and generally provide more useful information. Electro-acoustical transducers produce waves primarily of this type; other types generally cause little interference because of their lower speed.

The wave velocity depends upon the elastic properties and mass of the medium, and hence if the mass and velocity of wave propagation are known it is possible to assess the elastic properties. For an infinite, homogeneous, isotropic elastic medium, the compression wave velocity is given by:

$$V = \sqrt{\frac{K \cdot E_d}{\rho}} \tag{3.1}$$

where V = compression wave velocity (km/s)

$$K = \frac{(1 - v)}{(1 + v)(1 - 2v)}$$

E_d = dynamic modulus of elasticity (kN/mm^2)

ρ = density (kg/m^3)

and v = dynamic Poisson's ratio.

In this expression the value of K is relatively insensitive to variations of the dynamic Poisson's ratio v, and hence, provided that a reasonable estimate of this value and the density can be made, it is possible to compute E_d using a measured value of wave velocity V. Since v and ρ will vary little for mixes with natural aggregates, the relationship between velocity and dynamic elastic modulus may be expected to be reasonably consistent despite the fact that concrete is not necessarily the 'ideal' medium to which the mathematical relationship applies, as indicated in section 3.3.

3.2 Pulse velocity equipment and use

3.2.1 *Equipment*

The test equipment must provide a means of generating a pulse, transmitting this to the concrete, receiving and amplifying the pulse and measuring and displaying the time taken. The basic circuitry requirements are shown in Figure 3.1.

Repetitive voltage pulses are generated electronically and transformed into wave bursts of mechanical energy by the transmitting transducer, which must be coupled to the concrete surface through a suitable medium (see section 3.2.2). A similar receiving transducer is also coupled to the concrete at a known distance from the transmitter, and the mechanical energy converted back to electrical pulses of the same frequency. The electronic timing device measures the interval between the onset and reception of the pulse and this is displayed either on an oscilloscope or as a digital readout. The equipment must be able to measure the transit time to an accuracy of $\pm 1\%$. To ensure a sharp pulse onset, the electronic pulse to the transmitter must have rise time of less than one-quarter of its natural period. The repetition frequency of the pulse must be low enough to avoid interference between consecutive

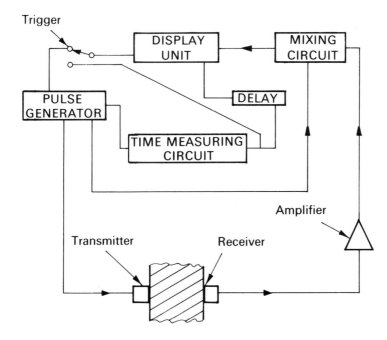

Figure 3.1 Typical UPV testing equipment.

pulses, and the performance must be maintained over a reasonable range of climatic and operating conditions.

Transducers with natural frequencies between 20 kHz and 150 kHz are the most suitable for use with concrete, and these may be of any type, although the piezo-electric crystal is most popular. Time measurement is based on detection of the compressive wave pulse, the first part of which may have only a very small amplitude. If an oscilloscope is used, the received pulse is amplified and the onset taken as the tangent point between the signal curve and the horizontal time-base line, whereas for digital instruments the pulse is amplified and shaped to trigger the timer from a point on the leading edge of a pulse.

A number of commercially produced instruments have become available in recent years which satisfy these requirements. The most popular of these are the V-meter produced in the USA (60) and the PUNDIT (Portable Ultrasonic Non-destructive Digital Indicating Tester) (61) produced in the United Kingdom. These have many similarities: both measure $180 \times 110 \times 160$ mm, weigh 3 kg, and have a digital display. Nickel-cadmium rechargeable batteries allow over nine hours continuous operation. Both also incorporate constant current charges to enable recharging from an a.c. mains supply, and may also be operated directly from the mains through a mains supply unit. For use in the laboratory an analogue unit can be added and this in turn can be connected to a recorder for continual experimental monitoring. Other available equipment incorporates an oscilloscope and permits amplitude monitoring.

Figure 3.2 shows the PUNDIT set up in the laboratory with 54 kHz transducers and a calibration reference bar. This steel bar has known characteristics and is used to set the zero of the instrument by means of a variable delay control unit each time it is used. The display is a four-digit liquid crystal and gives a direct transit time reading in microseconds. A wide range of transducers between 24 kHz and 200 kHz is available, although the 54 kHz and 82 kHz versions will normally be used for site or laboratory testing of concrete. Waterproof and even deep-sea versions of these transducers are available. An alternative form is the exponential probe transducer which makes a point contact (Figure 3.3), and offers operating advantages over flat transducers on rough or curved surfaces (see section 3.2.2.2). The equipment is robust and is provided with a carrying case for site use. Signal amplifiers are also available where long path lengths are involved on site, and the range of acceptable ambient temperatures of 0–45°C should cover most practical situations.

3.2.2 *Use*

Operation is relatively straightforward but requires great care if reliable results are to be obtained. One essential is good acoustical coupling between the concrete surface and the face of the transducer, and this is provided by

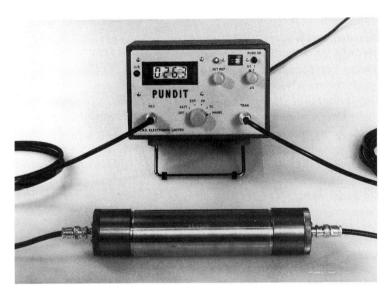

Figure 3.2 PUNDIT in laboratory (photograph by courtesy of C.N.S. Instruments Ltd).

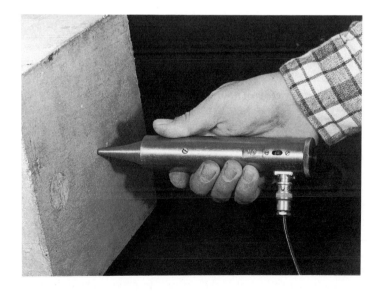

Figure 3.3 Exponential probe transducer.

a medium such as petroleum jelly, liquid soap or grease. Air pockets must be eliminated, and it is important that only a thin separating layer exists — any surplus must be squeezed out. A light medium, such as petroleum jelly or liquid soap, has been found to be the best for smooth surfaces, but a thicker grease is recommended for rougher surfaces which have not been cast against smooth shutters. If the surface is very rough or uneven, grinding or preparation with plaster of Paris or quick-setting mortar may be necessary to provide a smooth surface for transducer application. It is also important that readings are repeated by complete removal and re-application of transducers to obtain a minimum value for the transit time. Although the measuring equipment is claimed to be accurate to ± 0.1 microseconds, if a transit time accuracy of $\pm 1\%$ is to be achieved it may typically be necessary to obtain a reading to ± 0.7 microseconds over a 300 mm path length. This can only be achieved with careful attention to measurement technique, and any dubious readings should be repeated as necessary, with special attention to the elimination of any other source of vibration, however slight, during the test.

The path length must also be measured to an accuracy of $\pm 1\%$. This should present little difficulty with paths over about 500 mm, but for shorter paths it is recommended that calipers be used. The nominal member dimensions shown on drawings will seldom be adequate.

3.2.2.1 Transducer arrangement. There are three basic ways in which the transducers may be arranged, as shown in Figure 3.4. These are:

(i) Opposite faces (direct transmission)
(ii) Adjacent faces (semi-direct transmission)
(iii) Same face (indirect transmission).

Since the maximum pulse energy is transmitted at right angles to the face of the transmitter, the direct method is the most reliable from the point of view of transit time measurement. Also, the path is clearly defined and can be measured accurately, and this approach should be used wherever possible for assessing concrete quality. The semi-direct method can sometimes be used satisfactorily if the angle between the transducers is not too great, and if the path length is not too large. The sensitivity will be smaller, and if these requirements are not met it is possible that no clear signal will be received because of attenuation of the transmitted pulse. The path length is also less clearly defined due to the finite transducer size, but it is generally regarded as adequate to take this from centre to centre of transducer faces.

The indirect method is definitely the least satisfactory, since the received signal amplitude may be less than 3% of that for a comparable direct transmission. The received signal is dependent upon scattering of the pulse by discontinuities and is thus highly subject to errors. The pulse velocity will be predominantly influenced by the surface zone concrete, which may not

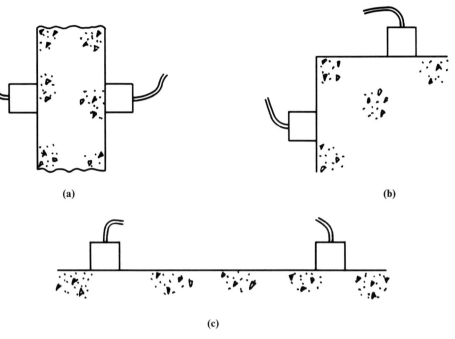

Figure 3.4 Types of reading. (a) Direct; (b) semi-direct; (c) indirect.

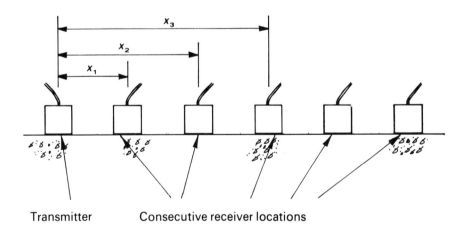

Transmitter   Consecutive receiver locations

Figure 3.5 Indirect reading — transducer arrangement.

be representative of the body, and the exact path length is uncertain. A special procedure is necessary to account for this lack of precision of path length, requiring a series of readings with the transmitter fixed and the receiver located at a series of fixed incremental points along a chosen radial line

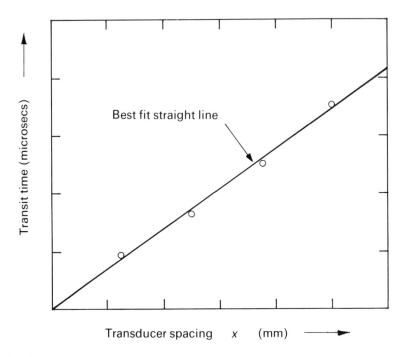

Figure 3.6 Indirect reading — results plot.

(Figure 3.5). The results are plotted (Figure 3.6) and the mean pulse velocity is given by the slope of the best straight line. If there is a discontinuity in this plot it is likely that either surface cracking or an inferior surface layer is present (see section 3.4). Unless measurements are being taken to detect such features, this method should be avoided if at all possible and only used where just one surface is available.

3.2.2.2 Transducer selection. The most commonly used transducers have a natural frequency of 54 kHz (Figure 3.2). They have a flat surface of 50 mm diameter, and thus good contact must be ensured over a considerable area. However, the use of a probe transducer making only point contact and normally requiring no surface treatment or couplant offers advantages. Time savings may be considerable and path length accuracy for indirect readings may be increased, but this type of transducer is unfortunately more sensitive to operator pressure. Receivers (as shown in Figure 3.3) have been found to operate satisfactorily in the field, but the signal power available from a transmitting transducer of this type is so low that its use is not normally practicable for site testing. The exponential probe receiver, which has a tip diameter of 6 mm, may also be useful on very rough surfaces where preparatory work might otherwise be necessary.

The only important factors which are likely to require the selection of an alternative transducer frequency relate to the dimensions of the member under test. Difficulties arise with small members as the medium under test cannot be considered as effectively infinite. This will occur when the path width is less than the wavelength λ. Since $\lambda = $ pulse velocity/frequency of vibration, it follows that the least lateral dimensions given in Table 3.1 should be satisfied. Aggregate size should similarly be less than λ to avoid reduction of wave energy and possible loss of signal at the receiver, although this will not normally be a problem. Although use of higher frequencies may reduce the maximum acceptable path length (10 m for 54 kHz to 3 m for 82 kHz), due to the lower energy output associated with the higher frequency, this problem can easily be overcome by the use of an inexpensive signal amplifier.

3.2.2.3 Equipment calibration. The time delay adjustment must be used to set the zero reading for the equipment before use, and this should also be regularly checked during and at the end of each period of use. Individual transducer and connecting lead characteristics will affect this adjustment, which is performed with the aid of a calibrated steel reference bar which has a transit time of around 25 μs. A reading through this bar (Figure 3.2) is taken in the normal way ensuring that only a very thin layer of couplant separates the bar and transducers. It is also recommended that the accuracy of transit time measurement of the equipment is checked by measurement on a second reference specimen, preferably with a transit time of around 100 μs.

3.3 Test calibration and interpretation of results

The basic problem is that the material under test consists of two separate constituents, matrix and aggregate, which have different elastic and strength properties. The relationship between pulse velocity and dynamic elastic modulus of the composite material measured by resonance tests on prisms is fairly reliable, as shown in Figure 3.7. Although this relationship is influenced by the value of dynamic Poisson's ratio, for most practical concretes made with natural aggregates the estimate of modulus of elasticity should be accurate within 10%.

Table 3.1 Minimum lateral path and maximum aggregate dimensions

Transducer frequency (kHz)	Minimum lateral path dimension or max. aggregate size (mm)	
	$V_c = 3.8$ km/s	$V_c = 4.6$ km/s
54	70	85
82	46	56
150	25	30

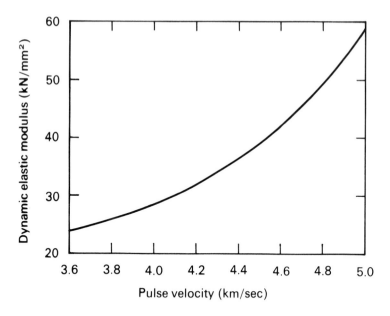

Figure 3.7 Pulse velocity vs. dynamic elastic modulus (based on ref. 58).

3.3.1 *Strength calibration*

The relationship between elastic modulus and strength of the composite material cannot be defined simply by consideration of the properties and proportions of individual constituents. This is because of the influence of aggregate particle shape, efficiency of the aggregate/matrix interface and variability of particle distribution, coupled with changes of matrix properties with age. Although some attempts have been made to represent this theoretically, the complexity of the interrelationships is such that experimental calibration for elastic modulus and pulse velocity/strength relationships is normally necessary. Aggregate may vary in type, shape, size and quantity, and the cement type, sand type, water/cement ratio and maturity are all important factors which influence the matrix properties and hence strength correlations. A pulse velocity/strength curve obtained with maturity as the only variable, for example, will differ from that obtained by varying the water/cement ratio for otherwise similar mixes, but testing at comparable maturities. Similarly, separate correlations will exist for varying aggregate types and proportions (Figures 3.8, 3.9 and 3.10) as well as for cement characteristics (62). This will include lightweight concretes (24) and special cements (63).

Strength calibration for a particular mix should normally be undertaken in the laboratory with due attention to the factors listed above. Pulse velocity readings are taken between both pairs of opposite cast faces of cubes of

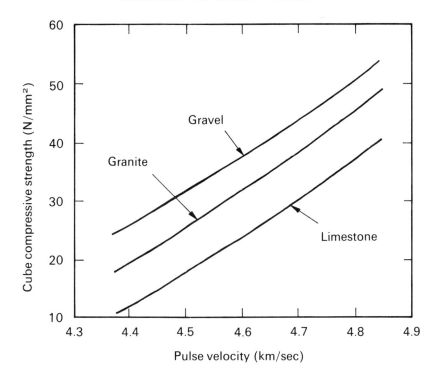

Figure 3.8 Effect of aggregate type (all concretes similar apart from aggregate type) (based on ref. 62).

known moisture condition, which are then crushed in the usual way. Ideally, at least 10 sets of three specimens should be used, covering as wide a range of strengths as possible, with the results of each group averaged. A minimum of three pulse velocity measurements should be taken for each cube, and each individual reading should be within 5% of the mean for that cube. Where this is not possible, cores cut from the hardened concrete may sometimes be used for calibration, although there is a danger that drilling damage may affect pulse velocity readings. Wherever possible readings should be taken at core locations prior to cutting. Provided that cores are greater than 100 mm in diameter, and that the ends are suitably prepared prior to test, it should be possible to obtain a good calibration, although this will usually cover only a restricted strength range. If it is necessary to use smaller diameter cores, high frequency transducers (section 3.2.2.2) may have to be used, and the accuracy of crushing strength will also be reduced (see section 5.3).

Although the precise relationship is affected by many variables, the curve may be expected to be of the general form

$$f_c = Ae^{BV}$$

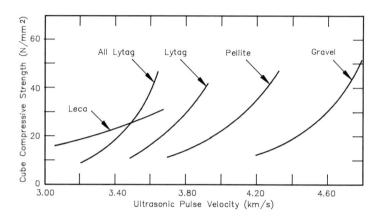

Figure 3.9 Comparison of lightweight and gravel aggregates (based on ref. 27).

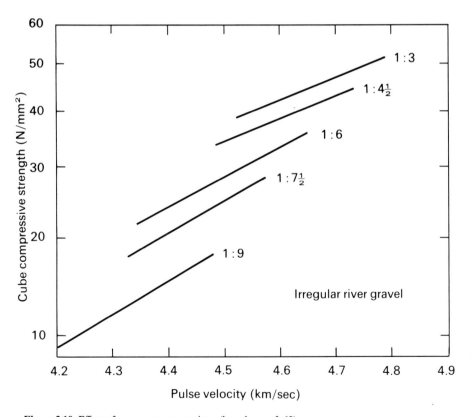

Figure 3.10 Effect of aggregate proportions (based on ref. 62).

where f_c = equivalent cube strength
  e = base of natural logarithms
  V = pulse velocity
and A and B are constants.

Hence a plot of log cube strength against pulse velocity is linear for a particular concrete. It is therefore possible to use a curve derived from reference specimens to extrapolate from a limited range of results from cores. Concrete made with lightweight aggregates is likely to give a lower pulse velocity at a given strength level. This is demonstrated in Figure 3.9, in which the effects of lightweight fines (All-Lytag) can also be seen. It should also be noted that for most lightweight aggregates there is likely to be reduced variability of measured values (24).

3.3.2 Practical factors influencing measured results
There are many factors relating to measurements made on in-situ concrete which may further influence results.

3.3.2.1 Temperature.
The operating temperature ranges to be expected in temperate climates are unlikely to have an important influence on pulse velocities, but if extreme temperatures are encountered their effect can be

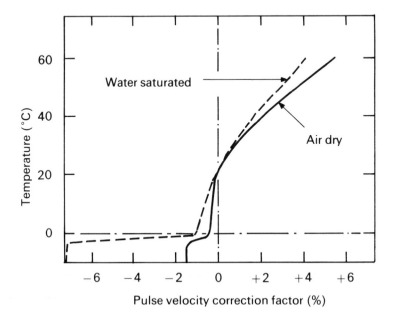

Figure 3.11 Effect of temperature (based on ref. 64).

estimated from Figure 3.11. These factors are based on work by Jones and Facaoaru (64) and reflect possible internal microcracking at high temperatures and the effects of water freezing within the concrete at very low temperatures.

3.3.2.2 Stress history.   It has been generally accepted that the pulse velocity of laboratory cubes is not significantly affected until a stress of approximately 50% of the crushing strength is reached. This has been confirmed by the authors (65) who have also shown from tests on beams that concrete subjected to flexural stress shows similar characteristics. At higher stress levels, an apparent reduction in pulse velocity is observed due to the formation of internal microcracks which will influence both path length and width.

It has been clearly shown that under service conditions in which stresses would not normally exceed $\frac{1}{3}$ cube strength the influence of compressive stress on pulse velocity is insignificant, and that pulse velocities for prestressed concrete members may be used with confidence. Only if a member has been seriously overstressed will pulse velocities be affected. Tensile stresses have been found to have a similarly insignificant effect, but potentially cracked regions should be treated with caution, even when measurements are parallel to cracks, since these may introduce path widths below acceptable limits.

3.3.2.3 Path length.   Pulse velocities are not generally influenced by path length provided that this is not excessively small, in which case the heterogeneous nature of the concrete may become important. Physical limitations of the time-measuring equipment may also introduce errors where short path lengths are involved. These effects are shown in Figure 3.12, in which a laboratory specimen has been incrementally reduced in length by sawing. BS 1881: Part 203 (58) recommends minimum path lengths of 100 mm and 150 mm for concrete with maximum aggregate sizes of 20 and 40 mm respectively. For unmoulded surfaces a minimum length of 150 mm should be adopted for direct, or 400 mm for indirect, readings.

There is evidence (50) that the measured velocity will decrease with increasing path length, and a typical reduction of 5% for a path length increase from approximately 3 m to 6 m is reported. This is because attenuation of the higher frequency pulse components results in a less clearly defined pulse onset. The characteristics of the measuring equipment are therefore an important factor. If there is any doubt about this, it is recommended that some verification tests are performed, although in most practical situations path length is unlikely to present a serious problem.

3.3.2.4 Moisture conditions.   The pulse velocity through saturated concrete may be up to 5% higher than through the same concrete in a dry condition, although the influence will be less for high-strength than for low-strength concretes. The effect of moisture condition on both pulse velocity and concrete strength is thus a further factor contributing to calibration difficulties, since

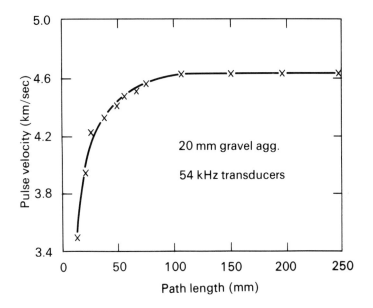

Figure 3.12 Effect of short path length (based on ref. 65).

the moisture content of concrete will generally decrease with age. A moist specimen shows a higher pulse velocity, but lower measured strength than a comparable dry specimen, so that drying out results in a decrease in measured pulse velocity relative to strength. The effect is well illustrated by the results in Figure 3.13 which relate to otherwise identical laboratory specimens, and demonstrates the need to correlate test cube moisture and structure moisture during strength calibration. It is thus apparent that strength correlation curves are of limited value for application to in-place concrete unless based on the appropriate maturity.

Tomsett (33) has presented an approach which permits calibration for 'actual' in-situ concrete strength to be obtained from a correlation based on standard control specimens. The relationship between specimens cured under different conditions is given as

$$\log_e \frac{f_1}{f_2} = kf_1(V_1 - V_2)$$

where f_1 is the strength of a 'standard' saturated specimen

f_2 is the 'actual' strength of the in-situ concrete

V_1 is the pulse velocity of the 'standard' saturated specimen

V_2 is the pulse velocity of the in-situ concrete

and k is a constant reflecting compaction control (a value of 0.015 is suggested

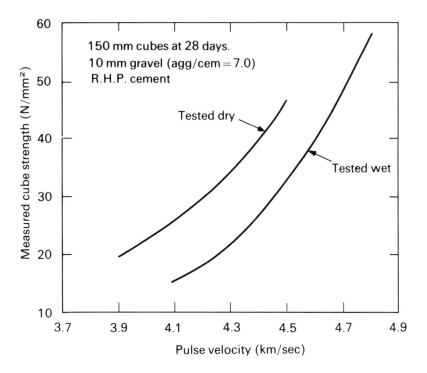

Figure 3.13 Effect of moisture conditions (based on ref. 65).

for normal structural concrete, or 0.025 if poorly compacted). This effect is illustrated by Figure 3.14, which is based on Tomsett's work. For any given curing conditions, it is possible to draw up a strength/pulse velocity relationship in this way, and similar members in a structure can be compared from a single correlation which may be assumed to have the same slope as the 'standard' saturated specimen relationship. This simple approach allows for both strength and moisture differences between in-situ concrete and control specimens. Swamy and Al-Hamed have also recommended a set of k values in a similar range, based on mix characteristics, and claim that these should enable in-situ strength estimation to within $\pm 10\%$ (66). However, a direct strength assessment of a typical reference specimen of in-situ concrete is still preferred if the relationship is to be used for other than comparative applications.

3.3.2.5 Reinforcement. Reinforcement, if present, should be avoided if at all possible, since considerable uncertainty is introduced by the higher velocity of pulses in steel coupled with possible compaction shortcomings in heavily reinforced regions. There will, however, often be circumstances in which it

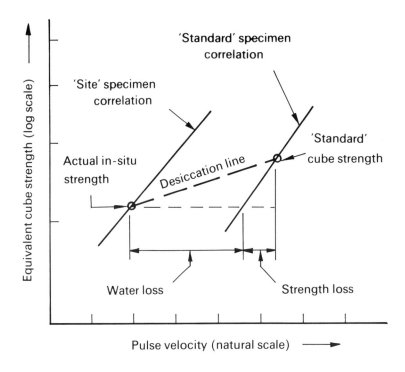

Figure 3.14 Desiccation line method (based on ref. 33).

is impossible to avoid reinforcing steel close to the pulse path, and corrections to the measured value will then be necessary. Corrections are not easy to establish, and the influence of the steel may dominate over the properties of the concrete so that confidence in estimated concrete pulse velocities will be reduced.

The pulse velocity in an infinite steel medium is close to 5.9 km/s, but this has been shown to reduce with bar diameter to as little as 5.1 km/s along the length of a 10 mm reinforcing bar in air (65). The velocity along a bar embedded in concrete is further affected by the velocity of pulses in the concrete and the condition of the bond between steel and concrete.

The apparent increase in pulse velocity through a concrete member depends upon the proximity of measurements to reinforcing bars, the diameter and number of bars and their orientation with respect to the propagation path. An increase will occur if the first pulse to arrive at the receiving transducer travels partly in concrete and partly in steel. Correction factors originally suggested by RILEM (67) assumed an average constant value of pulse velocity in steel and gave the maximum possible influence of the steel. The procedure adopted by BS 1881: Part 203 (58) is based on extensive experimental work by the authors (68) and takes bar diameter into account, yielding smaller corrections (see Figure 3.18). For practical purposes, with concrete pulse velocities of 4.0 km/s or above, 20-mm diameter bars running transversely to the pulse path will have no significant influence upon measured values, but bars larger than 6 mm diameter running along the path may have a significant effect.

There are two principal cases to be considered:

(i) *Axis of bars parallel to pulse path*
As shown in Figure 3.15, if a bar is sufficiently close to the path, the first wave to be received may have travelled along the bar for part of its journey. BS 1881: Part 203 suggests that a relationship of

$$V_c = \frac{2aV_s}{\sqrt{4a^2 + (TV_s - L)^2}} \quad \text{when} \quad V_s \geqslant V_c \tag{3.2}$$

is appropriate, where V_s = pulse velocity in steel bar
and V_c = pulse velocity in concrete

and that this effect disappears when

$$\frac{a}{L} > \frac{1}{2}\sqrt{\frac{V_s - V_e}{V_s + V_c}}$$

hence steel effects may be significant when $a/L < 0.15$ in high-quality concrete or <0.25 in low-quality material. The difficulty in applying eqn (3.2) lies in deciding the value of V_s. The equation can be expressed as

$$V_c = kV_m \tag{3.3}$$

where V_m is the measured apparent pulse velocity (L/T) in km/s, and k is the

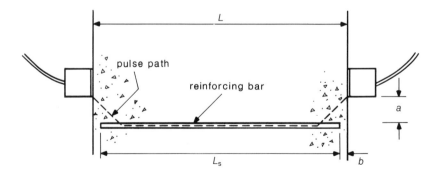

Figure 3.15 Reinforcement parallel to pulse path.

correction factor, given by

$$k = \gamma + 2\left(\frac{a}{L}\right)\sqrt{1 - \gamma^2} \qquad (3.4)$$

with

$$\gamma = \frac{V_c}{V_s}$$

The value of γ may be obtained from Figure 3.16 which has been plotted for a range of commonly occurring values of V_c and bar diameter, for a 54 kHz frequency. This may be substituted in eqn (3.4) (or Figure 3.17), to obtain a value of correction factor k to use in eqn (3.3). These equations are only valid where the offset a is greater than about twice the end cover to the bar b. Otherwise, pulses are likely to pass through the full length of the bar and

$$k = \gamma + 2\left(\frac{\sqrt{a^2 + b^2} - \gamma b}{L}\right) \qquad (3.5)$$

If the bar is directly in line with the transducers, $a = 0$ and the correction factor is given by

$$k = 1 - \frac{L_s}{L}(1 - \gamma) \qquad (3.6)$$

where L_s is the length of the bar (mm).

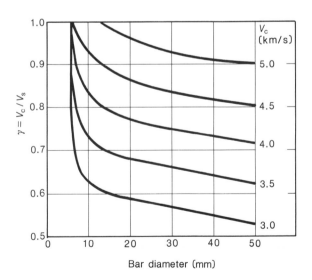

Figure 3.16 Relationship between bar diameter and velocity ratio for bars parallel to pulse path (based on ref. 58).

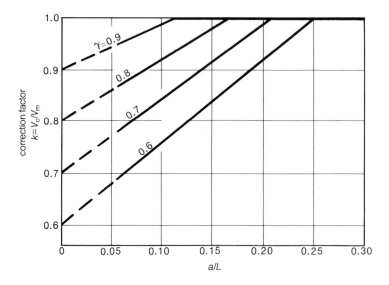

Figure 3.17 Correction factors for bars parallel to pulse path [a > 2b] (based on ref. 58).

An iterative procedure may be necessary to obtain a reliable estimate of V_c, and this is illustrated by an example in Appendix B. Estimates are likely to be accurate to within ±30% if there is good bond and no cracking of concrete in the test zone. Correction factors relating to a typical case of a bar in line with the transducers are shown in Figure 3.18, and compared with RILEM values which significantly overestimate steel effects for the smaller bar sizes.

Corrections must be treated with caution, especially since it is essentially the pulse through the concrete surrounding the bar which is being measured, rather than the body of the material. Complex bar configurations close to the test location will increase uncertainty.

(ii) *Axis of bars perpendicular to the pulse path*
For the situation shown in Figure 3.19a, if the total path length through steel across the bar diameters is L_s, the maximum possible steel effect is given by Figure 3.19b for varying bar diameters and concrete qualities, where V_c is the true velocity in the concrete.

In this case, the value of γ is used in eqn (3.6) to obtain the correction factor k. The effect on the bars on the pulse is complex, and the effective velocity in the steel is less than that along the axis of bars of similar size. Results for a typical case are shown in Figure 3.18, and the calculation procedure is illustrated in Appendix B.

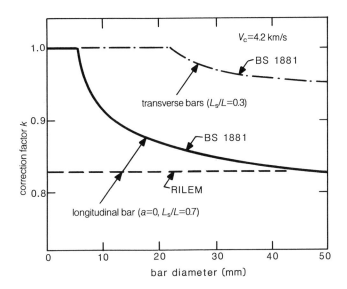

Figure 3.18 Typical correction factors.

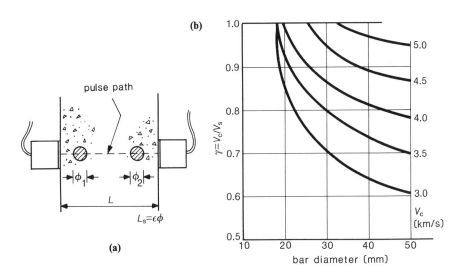

Figure 3.19 Reinforcement transverse to pulse path (based on ref. 58). (a) Path through transverse reinforcement. (b) Bar diameter/velocity ratio.

3.4 Applications

The applications of pulse velocity measurements are so wide-ranging that it would be impossible to list or describe them all. The principal applications are outlined below — the method can be used both in the laboratory and on site with equal success.

3.4.1 *Laboratory applications*

The principal laboratory applications lie in the monitoring of experiments which may be concerned either with material or structural behaviour. These include strength development or deterioration in specimens subjected to varying curing conditions, or to aggressive environments. The detection of the onset of micro-cracking may also be valuable during loading tests on structural members, although the method is relatively insensitive to very early cracking. For applications of this nature, the equipment is most effective if connected to a continuous recording device with the transducers clamped to the surface, thus removing the need for repeated application and associated operating errors.

3.4.2 *In-situ applications*

The wide-ranging and varied applications do not necessarily fall into distinct categories, but are grouped below according to practical aims and requirements.

3.4.2.1 Measurement of concrete uniformity. This is probably the most valuable and reliable application of the method in the field. There are many published reports of the use of ultrasonic pulse velocity surveys to examine the strength variations within members as discussed in Chapter 1. The statistical analysis of results, coupled with the production of pulse velocity contours for a structural member, may often also yield valuable information concerning variability of both material and construction standards. Readings should be taken on a regular grid over the member. A spacing of 1 m may be suitable for large uniform areas, but this should be reduced for small or variable units. Typical pulse velocity contours for a beam constructed from a number of batches are shown in Figure 3.20.

Tomsett (33) has suggested that for a single site-made unit constructed from a single load of concrete, a pulse velocity coefficient of variation of

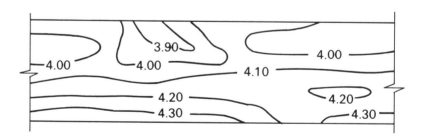

Figure 3.20 Typical pulse velocity beam contours (km/s).

1.5% would represent good construction standards, rising to 2.5% where several loads or a number of small units are involved. A corresponding typical value of 6–9% is also suggested for similar concrete throughout a whole structure. An analysis of this type may therefore be used as a measure of construction quality, and the location of substandard areas can be obtained from the 'contour' plot. The plotting of pulse velocity readings in histogram form may also prove valuable, since concrete of good quality will provide one clearly defined peak in the distribution (see section 1.6.2.1). Used in this way, ultrasonic pulse velocity testing could be regarded as a form of control testing, although the majority of practical cases in which this method has been used are related to suspected construction malpractice or deficiency of concrete supply. A survey of an existing structure will reveal and locate such features, which may not otherwise be detected. Although it is preferable to perform such surveys by means of direct readings across opposite faces of the member, Tomsett (69) has reported the successful use of indirect readings for comparison and determination of substandard areas of floor slabs.

Decisions concerning the seriousness of defects suggested by surveys of this type will normally require an estimate of concrete strength. As indicated in section 3.4.2.3, a reliable estimate of absolute strength is not possible unless a calibration is available. If the mean strength of the supply is known, the relationship $f_c = kV^4$ has been found satisfactory for estimating relative values over small ranges (33). Failing this, it will be necessary to resort to a more positive partially-destructive method, or core sampling, to obtain strength values, with the locations determined on the basis of the ultrasonic contour plot.

3.4.2.2 Detection of cracking and honeycombing. A valuable application of the ultrasonic pulse velocity techniques which does not require detailed correlation of pulse velocity with any other property of the material is in the detection of honeycombing and cracking. Since the pulse cannot travel through air, the presence of a crack or void on the path will increase the path length (as it goes around the flaw) and increase attenuation so that a longer transit time will be recorded. The apparent pulse velocity thus obtained will be lower than for the sound material. Since compression waves will travel through water it follows that this philosophy will apply only to cracks or voids which are not water-filled (shear waves will not pass through water). Tomsett (33) has examined this in detail and concluded that although water-filled cracks cannot be detected, water-filled voids will show a lower velocity than the surrounding concrete. Voids containing honeycombed concrete of low pulse velocity will behave similarly. The variation in pulse velocity due to experimental error is likely to be at least 2% notwithstanding variations in concrete properties, hence the size of a void must be sufficient to cause an increase in path length greater than 2% if it is to be detected. A given void is thus more difficult to detect as the path length increases, but the absolute minimum size of detectable defect will be set by the diameter of the transducer used.

In crack detection and measurement, even micro-cracking of concrete will be sufficient to disrupt the path taken by the pulses, and the authors (65) have shown that at compressive stresses in excess of 50% of the cube crushing strength, the measured pulse velocity may be expected to drop due to disruption of both path length and width. If the velocity for the sound concrete is known it is therefore possible to detect overstressing, or the onset of cracking may be detected by continual monitoring during load increase.

An estimate of crack depths may be obtained by the use of indirect surface readings as shown in Figure 3.21. In this case, where the transducers are equidistant from a known crack, if the pulse velocity through sound concrete is V km/s, then:

Path length without crack $= 2x$

Path length around crack $= 2\sqrt{x^2 + h^2}$

Surface travel time without crack $= \dfrac{2x}{V} = T_s$

Travel time around crack $= \dfrac{2\sqrt{x^2 + h^2}}{V} = T_c$

and it can be shown that

$$\text{crack depth, } h = x\sqrt{\left(\frac{T_c^2}{T_s^2} - 1\right)}.$$

An accuracy of $\pm 15\%$ can normally be expected, and this approach may be modified for applications to other situations as necessary.

Amon and Snell (70) have also described a number of case histories in which ultrasonic techniques have been used to monitor epoxy grout repairs to concrete based on the principle that poor bond or compaction will hinder the passage of pulses.

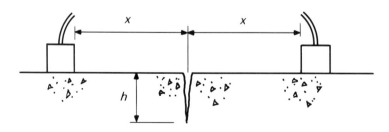

Figure 3.21 Crack depth measurement.

The location of honeycombing is best determined by the use of direct measurements through the suspect member, with readings taken on a regular grid. If the member is of constant thickness, a 'contour map' of transit times will readily show the location and extent of areas of poor compaction.

3.4.2.3 Strength estimation. Unless a suitable calibration curve can be obtained it is virtually impossible to predict the absolute strength of a body of in-situ concrete by pulse velocity measurements. Although it is possible to obtain reasonable correlations with both compressive and flexural strength in the laboratory, enabling the strength of comparable specimens to be estimated to $\pm 10\%$, the problems of relating these to in-situ concrete are considerable. If it is to be attempted, then the most reliable method is probably the use of cores to establish the calibration curve coupled with Tomsett's moisture correction. The authors (65) have suggested that if a reliable calibration chart is available, together with good testing conditions, it may be possible to achieve 95% confidence limits on a strength prediction of $\pm 20\%$ relating to a localized area of interest. Expected within-member variations are likely to reduce the corresponding accuracy of overall strength prediction of a member to the order of $\pm 10 \, \text{N/mm}^2$ at the $30 \, \text{N/mm}^2$ mean level. Accuracy decreases at higher strength levels, and estimates above $40 \, \text{N/mm}^2$ should be treated with great caution.

Although not perfect, there may be situations in which this approach may provide the only feasible method of in-situ strength estimation, and if this is necessary it is particularly important that especial attention is given to the relative moisture conditions of the calibration samples and the in-situ concrete. Failure to take account of this is most likely to cause an underestimate of in-place strength, and this underestimate may be substantial.

It is claimed (43) that significant improvements in accuracy can be obtained by combination with other techniques such as rebound hammer tests as described in Chapter 1, but this approach has never achieved popularity in the UK or USA.

3.4.2.4 Assessment of concrete deterioration. Ultrasonics are commonly used in attempting to define the extent and magnitude of deterioration resulting from fire, mechanical, frost or chemical attack. A general survey of the type described in section 3.4.2.1 will easily locate suspect areas (71), whilst a simple method for assessing the depth of fire or surface chemical attack has been suggested by Tomsett (33). In this approach it is assumed that the pulse velocity for the sound interior regions of the concrete can be obtained from unaffected areas, and that the damaged surface velocity is zero. A linear increase is assumed between the surface and interior to enable the depth to sound concrete to be calculated from a transit time measured across the damaged zone. For example, if a time T is obtained for a path length L including one damaged surface zone of thickness t, and the pulse velocity for sound concrete is V_c it can be shown that the thickness is given by

$$t = (TV_c - L)$$

Although this provides only a very rough estimate of damage depth, it is reported that the method has been found to give reasonable results in a number of fire damage investigations.

Where deterioration of the member is more general, it is possible that pulse velocities may reflect relative strengths either within or between members. There is a danger that elastic modulus, and hence pulse velocity, may not be affected to the same degree as strength and caution should therefore be exercised when using pulse velocities in this way.

Although it may be possible to develop laboratory calibrations for a mix subjected to a specific form of attack or deterioration, as was attempted when evaluating high alumina cement decomposition in the United Kingdom (72), absolute strength predictions of in-situ deteriorated concrete must be regarded as unreliable. In-situ comparison of similar members to identify those which are suspect for subsequent load testing, has however been carried out successfully in the course of a number of HAC investigations, and pulse velocities have been shown to be sensitive to the initiation and development of alkali–silica reaction (54, 73). This provides a relatively quick and cheap approach where a large number of precast units, for example, are involved. Long-term performance of concrete can also be monitored very successfully by conducting repetitive tests on the same element.

3.4.2.5 Measurement of layer thickness.

This is essentially a development of the indirect reading method which is based on the fact that as the path length increases the pulse will naturally tend to travel through concrete at an increasing depth below the surface. This is particularly appropriate for application to slabs in which a surface layer of different quality exists due to construction, weathering, or other damage such as fire. The procedure is exactly as described for obtaining an indirect measurement (section 3.2.2.1). When the transducers are close together the pulse will travel in the surface layer only, but at greater spacings the path will include the lower layer. This effect will be shown by a discontinuity in the plot of transit time v. transducer spacing, with the pulse velocities through the two layers having different slopes, as shown in Figure 3.22. The thickness t of the upper layer is related to the velocities V_1 and V_2, and the spacing x at which the discontinuity is observed, by the expression

$$t = \frac{x}{2}\sqrt{\frac{(V_2 - V_1)}{(V_2 + V_1)}}$$

Although this is most suitable for a distinct layer of uniform thickness, the value obtained can be at best only an estimate, and it must be borne in mind

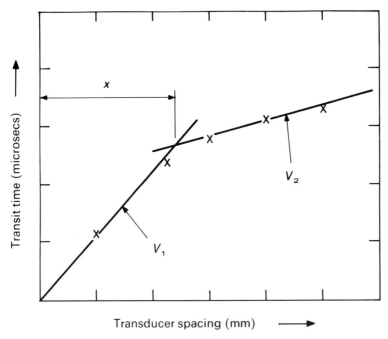

Figure 3.22 Layer thickness measurement.

that there will be a maximum thickness of layer that can be detected. Little information is available concerning the depth of penetration of indirect readings, and in view of the weakness of signal received using this method the results must be treated with care.

3.4.2.6 Measurement of elastic modulus. This is the property that can be measured with the greatest numerical accuracy. Values of pulse modulus can be calculated theoretically using an assumed value of Poisson's ratio to yield a value within $\pm 10\%$, or more commonly an estimate of dynamic modulus can be obtained from the reliable correlations with resonant frequency values. Whereas such measurements may be valuable in the laboratory when undertaking model testing, their usefulness on site is limited, although they may be used to provide an estimated static elastic modulus value for use in calculations relating to load tests.

3.4.2.7 Strength development monitoring. It has been well established that pulse velocity measurements will accurately monitor changes in the quality of the paste with time, and this may be usefully applied to the control of demoulding or stressing operations both in precasting works and on site. In this situation a specific pulse velocity/strength relationship for the mix, subject to the appropriate curing conditions, can be obtained and a safe acceptance level of pulse velocity established. In the same way, quality control of similar precast units may easily be undertaken and automated techniques incorporating amplitude assessment have been proposed (74).

3.5   Reliability and limitations

Ultrasonic pulse velocity measurement has been found to be a valuable and reliable method of examining the interior of a body of concrete in a truly non-destructive manner. Modern equipment is robust, reasonably cheap and easy to operate, and reliable even under site conditions; however, it cannot be overemphasized that operators must be well trained and aware of the factors affecting the readings. It is similarly essential that results are properly evaluated and intepreted by experienced engineers who are familiar with the technique. For comparative purposes the method has few limitations, other than when two opposite faces of a member are not available. The method provides the only readily available method of determining the extent of cracking within concrete; however, the use for detection of flaws within the concrete is not reliable when the concrete is wet.

Unfortunately, the least reliable application is for strength estimation of concrete. The factors influencing calibrations are so many that even under ideal conditions with a specific calibration it is unlikely that 95% confidence limits of better than $\pm 20\%$ can be achieved for an absolute strength prediction for in-place concrete. Although it is recognized that there may be some circumstances in which attempts must be made to use the method for strength prediction, this is not recommended. It is far better that attention is concentrated upon the use of the method for comparison of supposedly similar concrete, possibly in conjunction with some other form of testing, rather than attempt applications which are recognized as unreliable and which will therefore be regarded with scepticism.

4 Partially destructive strength tests

Considerable developments have taken place in recent years in methods which are intended to assess in-situ concrete strength, but cause some localized damage. This damage is sufficiently small to cause no loss in structural performance. All are surface zone tests which require access to only one exposed concrete face. Methods incorporate variations of the concepts of penetration resistance, pull-out, pull-off and break-off techniques which have been proposed over many years. Estimation of strength is by means of correlation charts which, in general, are not sensitive to as many variables as are rebound hammer or pulse-velocity testing. There are drawbacks in application and accuracy, which vary according to the method, but there are many circumstances in which these methods have been shown to be of considerable value. A key feature is that an estimate of strength is immediately available, compared with delays of several days for core testing, and although accuracy may not be as good, the testing is considerably less disruptive and damaging.

The best-established of these methods are covered by American and other national standards, and are incorporated in BS 1881: Part 207 (75) and a report by ACI Committee 228 (37). The choice of method for particular circumstances will depend largely upon whether the testing is preplanned before casting, together with practical factors such as access, cost, speed and prior knowledge of the concrete involved. The tests can be used only where making good of surfaces is acceptable.

4.1 Penetration resistance testing

The technique of firing steel nails or bolts into a concrete surface to provide fixings is well established, and it is known that the depth of penetration is influenced by the strength of the concrete. A strength determination method based on this approach, using a specially designed bolt and standardized explosive cartridge, was developed in the USA during the mid 1960s and is known as the Windsor probe test (76). It has gained popularity in the USA and Canada, especially for monitoring strength development on site, and is the subject of ASTM C803 (77). Many authorities in North America regard it as equivalent to site cores, and in some cases it is accepted in lieu of control cylinders for compliance testing. Use outside North America has been limited, but the equipment is readily available and the method is included in BS 1881: Part 207.

Although it is difficult to relate theoretically the depth of penetration of the bolt to the concrete strength, consistent empirical relationships can be found that are virtually unaffected by operator technique. The method is a form of hardness testing and the measurements will relate only to the quality of concrete near the surface, but it is claimed that it is the zone between approximately 25 and 75 mm below the surface which influences the penetration. The depth is considerably greater than for rebound or any other established 'surface zone' tests.

A smaller-scale method has also been proposed (78) in which a spring-loaded hammer drives a small pin into the concrete surface to a depth of between 4–8 mm. This pin penetration test is primarily intended for determination of in-situ concrete strength to permit formwork stripping.

4.1.1   *Windsor probe*

4.1.1.1   Test equipment and operation.   The bolt or probe which is fired into the concrete (Figure 4.1) is of a hardened steel alloy. The principal features are a blunt conical end to punch through the matrix and aggregate near the surface, and a shoulder to improve adhesion to the compressed concrete and ensure a firm embedment.

The probes are generally 6.35 mm in diameter and 79.5 mm in length, but larger-diameter bolts (7.94 mm) are available for testing lightweight concretes.

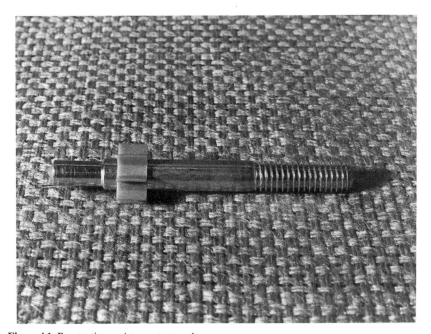

Figure 4.1 Penetration resistance test probe.

Figure 4.2 Driver in use.

A steel firing head is screwed on to the threaded end of the bolt and the plastic guide locates the probe within the muzzle of the driver from which it is fired. The driver, which is shown in operation in Figure 4.2, utilizes a carefully standardized powder cartridge. This imparts a constant amount of energy to the probe irrespective of firing orientation, and produces a velocity of 183 m/s which does not vary by more than $\pm 1\%$. The power level can be reduced when dealing with low strength concretes simply by locating the probe at a fixed position within the driver barrel. The driver is pressed firmly against a steel locating plate held on the surface of the concrete which releases a safety catch and permits firing when the trigger is pulled. After firing, the driver head and locating plate are removed and any surface debris around the probe is scraped or brushed away to give a level surface. A flat steel plate is placed on this surface, and a steel cap screwed onto the probe to enable the exposed length to be measured to the nearest 0.5 mm with a spring-loaded calibrated depth gauge, as in Figure 4.3.

Probe penetrations may be measured individually as described, or alternatively the probes may be measured in groups of three using a triangular template with the probes at 177 mm centres. In this case, a system of triangular measuring plates is used which will provide one averaged reading of exposed length for the group of probes. This approach may mask inconsistencies between individual probes, and it is preferable to measure each probe individually. The measured average value of exposed probe length may then

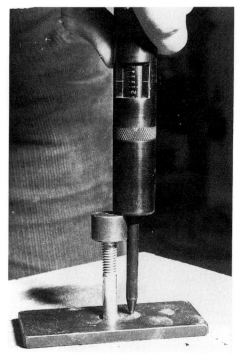

Figure 4.3 Height measurement.

be directly related to the concrete strength by means of appropriate calibration tables or charts.

It is important to recognize that in the UK it is necessary to comply with the requirements of BS 4078: Part 1 (79) concerning the use of powder-actuated driving units, as well as a range of Health and Safety Acts which are listed in BS 1881: Part 207 (75). These restrictions may limit the use of the technique in some situations.

4.1.1.2 Procedure. Individual probes may be affected by particularly strong aggregate particles near the surface, and it is thus recommended that at least three tests are made and averaged to provide a result. If the range of a group of three tests exceeds 5 mm, a further test should be made and the extreme value discarded. Although slight surface roughness is not important, surfaces coarser than a broom finish should be ground smooth prior to test, and the probe must always be driven perpendicular to the surface.

Where the expected cube strength of the concrete is less than 26 N/mm², the 'low power' setting should be used, but for higher strengths, this penetration may not be sufficient to ensure firm embedment of the probe, and the 'standard power' setting is necessary. If probes will not remain fixed in very high strength concrete it may be possible to measure directly the depth of hole formed, after cleaning, and subtract this from the probe length. It should be noted, however, that this does not conform to BS 1881: Part 207 requirements. The manufacturers of the system recommend that a minimum edge distance of 100 mm should be maintained (75 mm for low power) but the authors' experience suggests that these values may not always be sufficient to prevent splitting. Probes should also be at least 175 mm apart to avoid overlapping of zones of influence.

BS 1881: Part 207 recommends a 150 mm edge distance and 200 mm minimum spacing, with the added restriction that a test should not be located within 50 mm of a reinforcing bar. A minimum concrete element thickness of 150 mm is also recommended.

Aggregate hardness is an important factor in relating penetration to strength, and it may therefore be necessary to determine its value. This is assessed on the basis of the Mohs' hardness scale, which is a system for classifying minerals in terms of hardness into ten groups. Group 10 is the hardest and Group 1 the softest, thus any mineral will scratch another from a group lower than itself. Testing consists of scratching the surface of a typical aggregate particle with minerals of known hardness from a test kit; the hardest is used first, then the others in order of decreasing hardness until the scratch mark will wipe off. The first scratch that can be wiped off represents the Mohs' classification for the aggregate.

4.1.1.3 Theory, calibration and interpretation. A convincing theoretical description of the penetration of a concrete mass by a probe is not available, since there is little doubt that a complex combination of compressive, tensile, shear and friction forces must exist. The manufacturers of the Windsor probe equipment have suggested that penetration is resisted by a subsurface compressive compaction bulb as shown in Figure 4.4. The surface concrete will crush under the tip of the probe, and the shock waves associated with the impact will cause fracture lines, and hence surface spalling, adjacent to the probe as it penetrates the body of concrete. The energy required to cause this spalling, or to break pieces of aggregate, is a low percentage of the total energy of a driven probe, and will therefore have a small effect upon the depth of penetration. Penetration will continue, with cracks not necessarily reaching the surface and eventually ceasing to form as the stress drops. Energy is absorbed by the continuous crushing at the point, by surface friction and by compression of the bulb of contained concrete. It is this latter effect which prevents rebound of the probe, and it is claimed that the bulb, and depth of penetration, will be inversely proportional to the compressive strength. Data are not currently available to support these proposals, which must be regarded as rather simplistic, although the concept of the measured property relating to concrete below rather than at the surface seems reasonable.

Although it may theoretically be possible to undertake calculations based on the absorption of the kinetic energy of the probe, this would be difficult,

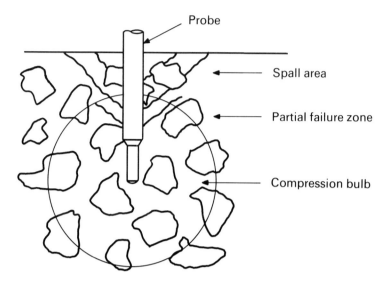

Probe

Spall area

Partial failure zone

Compression bulb

Figure 4.4 Compaction bulb.

and it is very much easier to establish empirical relationships between penetration and strength.

Calibration is hampered by the minimum edge distance requirement which prevents splitting. Although it may be possible to use standard 150 mm cubes or cylinders for tests at low power, the specimen must be securely held during the test. A holding jig for cylinders is available from the Windsor probe manufacturers, and cubes are most conveniently clamped in a compression-testing machine, although no data concerning the infuence of applied compressive strength are available. It is recommended by Malhotra (50) that groups of at least six specimens from the same batch are used, with three tested in compression and three each with one probe test, and the results averaged to produce one point on the calibration graph. Malhotra has also shown that the reduction in measured compressive strength of cylinders which have been previously probed may be up to 17.5%, and such specimens cannot therefore be tested in compression for calibration purposes.

Where the cube strength of the concrete is greater than 26 N/mm^2 it is necessary to use a combination of cubes or cylinders for compression testing and larger slab or beam specimens from the same batch for probing. The size of such specimens is unimportant provided that they are large enough to accommodate at least three probes which satisfy the minimum edge distance and spacing requirements. These test specimens must be similarly compacted, and all should be cured together. In such situations, the use of ultrasonic pulse velocity measurements to compare concrete quality between specimens would be valuable. This approach has been used by the authors

(80) in an investigation in which 1000 × 250 × 150 mm beams were used for probing, and 100 mm cubes for compression testing, and it was found that the beam concrete was between 10 and 20% lower in strength than the concrete in the cubes. Since calibrations will normally relate to actual concrete strength it is also important that the moisture conditions of the specimens are similar. Figure 4.5 shows a typical calibration chart obtained in this way with a strength range obtained by water/cement ratio and age variations. Relationships between penetration and strength for the two different power levels are not easily related, and it is therefore necessary to produce calibration charts for each experimentally.

The manufacturers of the test equipment provide calibration tables (76) in which aggregate hardness is taken as the only variable influencing the penetration/strength relationship. It is clear from the authors' work and from reported experience in the USA (81) that this is not the case, and that aggregate type can also have a large influence. It is understood that the manufacturer's tables are based on crushed rock, but for rounded gravels the crushing strength may be lower than suggested by probe results. It is to be expected that bond differences at the aggregate/matrix interface due to aggregate

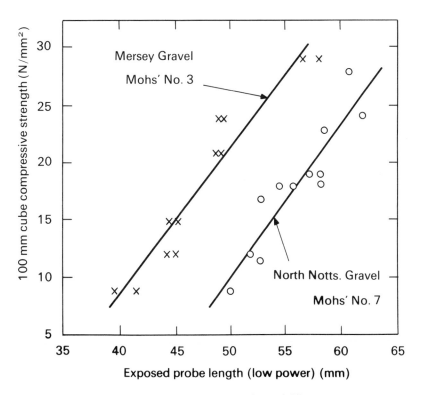

Figure 4.5 Typical low power strength calibration (based on ref. 80).

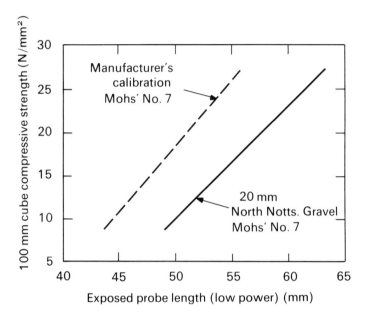

Figure 4.6 Comparison of calibrations (based on refs 76 and 80).

surface characteristics may affect penetration resistance and crushing strength. Nevertheless, the extent of the calibration discrepancy which may be attributed to this, as indicated in Figure 4.6, is disturbing.

Calibrations from a number of sources are compared in Figure 4.7. It appears that moisture condition, aggregate size (up to 50 mm) and aggregate proportions all have effects which are small in relation to aggregate hardness and type. Swamy and Al-Hamed (82) have also suggested that curing conditions and age are important, with differing penetration/strength relationships for old and new concrete. It is essential therefore that appropriate calibration charts should be developed for the particular aggregate type involved in any practical application of the method, and this requirement has also been confirmed for lightweight concretes (24).

4.1.1.4 Reliability, limitations and applications. The test is not greatly affected by operator technique, although verticality of the bolt relative to the surface is obviously important and a safety device in the driver prevents firing if alignment is poor. It is claimed that an average coefficient of variation for a series of groups of three readings on similar concrete of the order of 5% may be expected, and that a correlation coefficient of greater than 0.98 can be achieved for a linear calibration relationship for a single mix. Field tests by the authors on motorway deck slabs have also yielded a similar coefficient of variation of probe results over areas involving several truck loads of concrete. It is also apparent from Figure 4.5 that 95% limits of

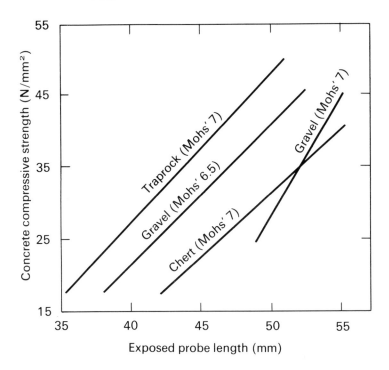

Figure 4.7 Influence of aggregate type and proportions (based on ref. 50).

about $\pm 20\%$ on predicted strengths may be possible for a single set of three probes, given adequate calibration charts. Difficulty may be encountered in predicting strengths in the range 25–50 N/mm^2 at ages greater than one year (82), and in the authors' experience the method cannot be reliably used for strengths below about 10 N/mm^2. Results for lightweight concrete (24) suggest that accuracy levels may be reduced when lightweight fines are present. It is to be expected that aggregate size will influence the scatter of individual probe readings, but at present insufficient data are available to assess the effect of this on strength prediction accuracies, although a 50 mm maximum size is recommended. Similarly, the effect of reinforcement adjacent to the probe is uncertain, and a minimum clearance of 50 mm should be allowed between probes and reinforcing bars.

The principal physical limitation of this method is caused by the need for adequate edge distances and probe spacings together with a member thickness of at least twice the anticipated penetration. After measurement the probe can be extracted, leaving a conical damage zone (Figure 4.8) which must be made good. There is the additional danger of splitting of the member if it does not comply with the minimum recommended dimensions. Expense is a further consideration with relatively high cost equipment and recurrent costs and the safety aspects outlined previously cannot be ignored.

Figure 4.8 Surface damage caused by probe removal.

The limitations outlined above mean that although probe measurement takes place at a greater depth within the concrete than rebound hammer measurements, penetration tests are unlikely to replace rebound tests except where the latter are clearly unsatisfactory. Probes cannot examine the interior of a member in the same way as ultrasonics and the method causes damage that must be repaired. However, probes do offer the advantage of requiring only one surface and fewer calibration variables. In relation to cores, however, probes provide easier testing methods, speedy results and accuracy of strength estimation comparable to small-diameter specimens. Although the accuracies of large-diameter cores cannot be matched, it is likely that probing may be used as an alternative to cores in some circumstances.

In the USA there is a trend towards in-place compliance testing, especially in relation to post-tensioning. Since the most reliable application of the penetration method lies in comparison of similar concrete where specific calibration charts can be obtained, a number of applications of this nature has been reported. It appears that the advantages of speed and simplicity, together with the ability to drive probes through timber or even thin steel formwork without influence, outweigh the cost. Details of acceptable thicknesses are unfortunately not available, but in such circumstances decisions would normally be based on previously established 'go/no go' limits for measured penetration.

Other applications include the detection of substandard members or areas of mature concrete, and this method is particularly appropriate for large walls or slabs having only one exposed surface free of finishes. Investigations of this type have been successfully performed by the authors on highway bridge deck slabs. Probing was carried out on the deck soffit from a small mobile hydraulic platform while the road above was in normal use, and in one such investigation a total of 18 sets of probes were placed by one operator in a period of six hours. The speed of operation, together with the immediate availability of results, means that many more tests can be made than if cores were being taken, and test locations can be determined in the light of the results obtained. This is particularly valuable when attempting to define the location and extent of substandard concrete.

Whether or not the test will be of significant value in the strength assessment of 'unknown' concrete is uncertain, but it is clear that results based solely on aggregate hardness are inadequate. It may be, however, that as more results are made available it will be possible to increase the confidence with which the method may be extended beyond comparative situations.

4.1.2 Pin penetration test

Nasser and Al-Manaseer (78) have developed this new method, aimed at determining formwork stripping times. The apparatus consists of a spring-loaded hammer which can grip a pin of 30.5 mm length and 3.56 mm diameter with the tip machined at an angle of 22.5°. The spring is compressed by pressing the hammer against the concrete surface, and is released by a trigger causing the pin and the attaached shaft and hammer to impact the concrete surface with an energy of about 108 Nm. The depth of the hole created is measured with a dial gauge device after cleaning with an air blower.

Calibration testing with gravel and lightweight concretes with cylinder strengths between 3.1 and 24.1 N/mm^2 has shown linear relationships between penetration and compressive strength which have good correlation coefficients and are very close to each other (78). It is suggested that for practical purposes these can be combined. The same authors have subsequently compared the performance with a range of other test methods (81) and shown that the method compared well in terms of correlation accuracy being the only method not requiring separate calibration for lightweight concrete.

It is suggested that a reading should be taken as the average of the best five of a group of seven tests to allow for local influences. The principal advantages of the method seem to be its speed, simplicity, low cost and low level of damage. The depth of penetration is unlikely to exceed 8 mm, hence reinforcement poses no problem. Results at present are limited, both in terms of strength range and aggregate and mix types. Features such as carbonation and temperature have yet to be examined in detail, but Shoya et al. (83) have provided some data suggesting coefficients of variation up to 18% and indicating difficulties of strength prediction with deteriorated or carbonated surfaces. The method seems to offer potential, however, worthy of further investigation.

4.2   Pull-out testing

The concept of measuring the force needed to pull a bolt or some similar device from a concrete surface has been under examination for many years. Proposed tests fall into two basic categories; those which involve an insert which is cast into the concrete, and those which offer the greater flexibility of an insert fixed into a hole drilled into the hardened concrete. Cast-in methods must be preplanned and will thus be of value only in testing for specification compliance, whereas drilled-hole methods will be more appropriate for field surveys of mature concrete. In both cases, the value of the test depends upon the ability to relate pull-out forces to concrete strengths and a particularly valuable feature is that this relationship is relatively unaffected by mix characteristics and curing history. Although the results will relate to the surface zone only, the approach offers the advantage of providing a more direct measure of strength and at a greater depth than surface hardness testing by rebound methods, but still requires only one exposed surface. Procedures have recently been reviewed in detail by Carino (84).

4.2.1   Cast-in methods

Reports were first published in the USA and USSR in the late 1930s describing tests in which a cast-in bolt is pulled from the concrete. These methods do not appear to have become popular, and it was not until 30 years later that practically feasible tests were developed. Two basic methods, both of which require a threaded insert which is fixed to the shuttering prior to concreting, have emerged. A bolt is then screwed into the insert and pulled hydraulically against a circular reaction ring. The principal difference between the two systems, developed in Denmark and Canada respectively, lies in the shape of insert and loading technique. In both cases a cone of concrete is 'pulled out' with the bolt, and the force required to achieve this is translated to compressive strength by the use of an empirical calibration.

4.2.1.1   The Lok-test.   This approach, developed at the Danish Technical University in the late 1960s, has gained popularity in Scandinavia and is now accepted by a number of public agencies in Denmark as equivalent to cylinders for acceptance testing (1).

The insert (Figure 4.9) consists of a steel sleeve which is attached to a 25 mm diameter, 8 mm thick anchor plate located at a depth of 25 mm below the concrete surface (85). The sleeve is normally screwed to the shuttering, or fixed to a plastic buoyancy cup where slabs are to be tested. This is later removed and replaced by a rod of 7.2 mm diameter which is screwed into the anchor plate and coupled to a tension jack. The whole assembly is

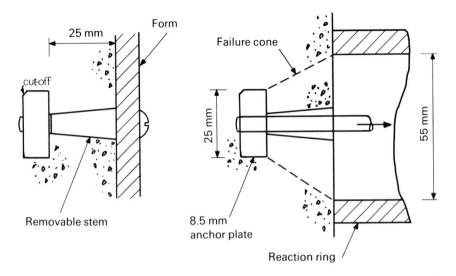

Figure 4.9 Lok-test insert.

pre-coated to prevent bonding to the concrete, and rotation of the plate is prevented by the 'cut-off'. A special extension device is also available to permit tests at greater depth if required. Load is applied to the pull-bolt by means of a portable hand-operated hydraulic jack with a reaction ring of 55 mm diameter. This equipment (Figure 4.10) is compact, with a weight of less than 5 kg.

The loading equipment can determine the force required to cause failure by pulling the disc, and a range of jacks is available to cover all practical concrete strengths including 'high-strength' concretes. The load is measured with an accuracy of $\pm 2\%$ over normal operating temperatures, and a precision valve system combined with a friction coupling ensures a constant loading rate of 30 ± 10 kN/min. Electronic digital reading apparatus with data storage facilities has recently become available. Load is released as soon as a peak is reached, leaving only a fine circular crack on the concrete surface. Calibration charts as those provided by Petersen (1, 86) (Figure 4.11) or the authors (87) are then used to estimate the compressive strength of the concrete.

BS 1881: Part 207 recommends that the centres of test positions should be separated by at least eight times the insert head diameter, and that minimum edge distances should be four diameters. An element thickness of at least four insert head diameters is needed and tests should be located so that there is no reinforcing steel within one bar diameter (or maximum aggregate size if greater) of the expected conic fracture surface. A minimum of four tests is recommended to provide a result for a given location.

The geometric configuration indicated in Figure 4.9 ensures that the failure

Figure 4.10 Lok-test equipment (photograph by courtesy of Lok-test Aps.).

surface is conical and at an angle of approximately 31° to the axis of applied tensile force. This is close to the angle of friction of concrete, which is generally assumed to be 37°, and extensive theoretical work has shown that this produces the most reliable measure of compressive strength. Plasticity theory for concrete using a modified Coulomb's failure criterion indicates that where the failure angle and friction angle are equal, the pull-out force is proportional to compressive strength. Finite element analyses of the failure mechanism (88) have indicated that failure is initiated by crushing, rather than cracking, of the concrete. It is suggested that a narrow symmetrical band of compressive forces runs between the cast-in disc and the reaction tube on the surface. Further theoretical (89) and experimental (90) research effort has been devoted to attempts to explain the failure mechanisms, and differing views remain. These primarily concern the relative importance of compressive crushing and aggregate interlock effects following initial circumferential cracking which is generally agreed to be fully developed at about 65% of the final pull-out load.

Stone and Giza (91) have examined in detail the effect of changes in geometry and the test assembly and the effect of concrete aggregate properties on the reliability of the pull-out test. For concrete with cylinder strengths in the 14–17 N/mm² range they have concluded that pull-out force decreases with increasing apex angle, but that there is no change in variability for apex angles between 54° and 86°, although scatter increases rapidly for lower angles. As would be expected, pull-out load increases with depth of embedment and it is confirmed that it is not affected by aggregate type or size. Variability was, however, shown to be greater for 19-mm aggregate than for smaller

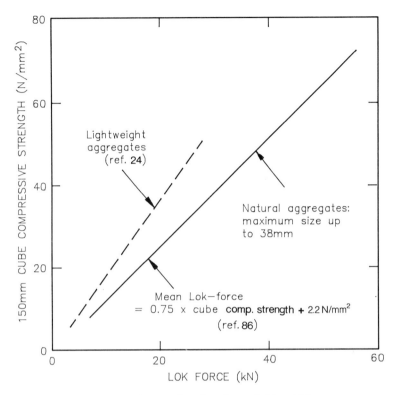

Figure 4.11 Typical Lok-test calibration chart (based on refs 24 and 86).

sizes, and mortar specimens showed less variation and lower failure loads than corresponding specimens containing natural aggregate. Low variability was similarly found for lightweight aggregate concrete.

The reliability of the method is reported to be good, with correlation coefficients for laboratory calibrations of about 0.96 on straight line relationships, and a corresponding coefficient of variation of about 7%. Comparison with rebound hammer and ultrasonic pulse velocity strength calibrations shows that the slope is much steeper, hence this test is much more sensitive to strength variations. An important feature of this approach is the independence of the calibration of features such as water/cement ratio, curing, cement type and natural aggregate properties (up to 38 mm maximum size). Strength calibration is thus more dependable than for most other non-destructive or partially-destructive methods and generalized correlations may be acceptable with prediction accuracies of the order of ±20%. However, for large projects it is recommended that a specific calibration is developed for the concrete actually to be used in which case 95% confidence limits of ±10% may be possible. It should also be noted that artificial lightweight aggregates are

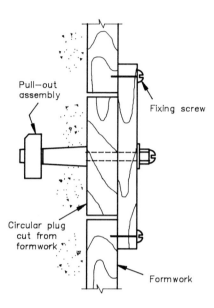

Pull—out
assembly

Fixing screw

Circular plug
cut from
formwork

Formwork

Figure 4.12 Arrangement for formwork stripping time tests.

likely to require specific calibration as illustrated in Figure 4.11 (24) which shows a reduced value of pull-out force for a particular compressive strength level. The two principal limitations are preplanned usage (although the Capo test, section 4.2.2.3, overcomes this), and the surface zone nature of the test. The test equipment can be obtained in a convenient briefcase kit form containing all the necessary ancillary items, although the cost is relatively high.

Bickley (92) indicates that the use of this approach is growing rapidly in North America, especially for determination of form stripping times, and provides illustrative examples of statistical analysis in relation to specification criteria. There seem to be few practical problems associated with in-situ usage, and an arrangement such as that shown in Figure 4.12 may be convenient in this situation. The technique has been shown to be particularly suitable for testing at very low concrete strengths and at early ages as illustrated by the authors' results in Figure 4.13. Other applications include determination of stressing time in post-tensioned construction, whilst in Denmark the approach is accepted as a standard in-situ strength determination method and may form the basis of specification compliance assessment (1). Its use in many parts of the world for in-situ strength monitoring has been considerable, and this is likely to spread in the future.

4.2.1.2 North American pull-out methods. In the early 1970s Richards published data from tests made using equipment of his design (84), the basic

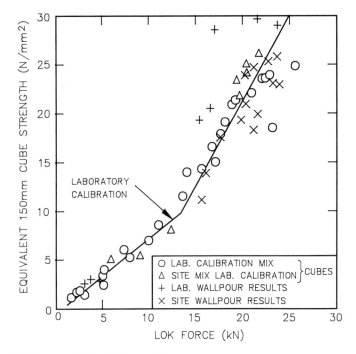

Figure 4.13 Low strength Lok-test correlation.

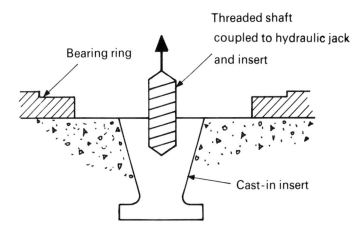

Figure 4.14 'American' insert.

form of which is shown in Figure 4.14. During subsequent years a number of test programmes were reported in the United States and Canada using this approach and other comparable test assemblies. These were sufficient to confirm the potential value of the method, and an American (93) standard has subsequently been developed. ASTM C900 allows considerable latitude in the details of the test assembly while specifying ranges of basic relative dimensions. It is intended that a hydraulic ram is used for load application, which should be at a uniform rate over a period of approximately two minutes. The depth of test may be greater than that of the Lok-test (section 4.2.1.1) athough this equipment does satisfy the requirements of both American and Canadian standards. Indeed, recent reports suggest that use of the commercially available Lok-test system dominates in these countries in preference to other versions of the method.

The failure surface will be less precisely defined than with the Lok-test because of the range of allowable dimensions, but Richards (94) suggests that an apex angle of 67° is most satisfactory. Although little theoretical work has been published relating to this type of insert it is likely that mechanisms will occur which are similar to those for the Lok-test. Presentation of results, however, according to ASTM C900 (93) should be in the form of a pull-out strength (f_p) calculated from the ratio of pull-out force to the failure surface area

$$f_\mathrm{p} = \frac{F}{A}$$

where F = force on ram
and A = failure surface area.
A may be calculated from

$$A = \frac{\pi}{4}(d_3 + d_2)(4h^2 + (d_3 - d_2)^2)^{1/2}$$

where d_2 = diameter of pull-out insert head
d_3 = inside diameter of reaction ring
h = distance from insert head to the surface.

The published numerical data relating to this method are not extensive, but correlation coefficients of 0.99 are claimed (94) between pull-out and core strength, with a relationship of pull-out strength = 0.21 × core strength. Coefficients of variation on testing would appear to be influenced by the nature of the insert, but Richards (94) also reports values as low as 7% for precision turned inserts.

Applications are obviously limited to preplanned situations and will be similar to those discussed for the Lok-test; similar limitations will also apply.

4.2.2 *Drilled-hole methods*

These offer the great advantage that use need not be preplanned. Early proposals from the USSR involved bolts grouted into the holes, but more recently two alternative methods have been developed and are both commanding interest. In 1977 the use of expanding wedge anchor bolts was proposed by Chabowski and Bryden-Smith (95), working for the Building Research

Establishment. Their technique was initially developed for use with pretensioned high alumina cement concrete beams and is known as the internal fracture test. This work has subsequently been extended to Portland cement concretes (96), and the authors have suggested that an alternative loading technique offers greater reliability (97). In Denmark, work on the Lok-test (see 4.2.1.1) has been extended (1) to produce the Capo test (cut and pull-out) in which an expanding ring is fixed into an under-reamed groove, producing a similar pull-out configuration to that used for the Lok-test.

Research in Canada (98) has also considered drilled-hole methods incorporating split sleeve assemblies, as well as reviving the concept of bolts set into hardened concrete using epoxy. This suggests that, despite practical problems and high test variability, both of these approaches are worthy of future development. An expanding sleeve device has also been proposed in the UK (99) and called ESCOT. There is little doubt that if a reliable drilled-hole pull-out approach could be established, it would be extremely valuable for in-situ concrete strength assessment, especially when the concrete mix details are unknown.

4.2.2.1 Internal fracture tests. The basic procedures for this method (95) are as follows. A hole is drilled 30–35 mm deep into the concrete using a roto-hammer drill with a nominal 6 mm bit. The hole is then cleared of dust with an air blower and a 6 mm wedge anchor bolt with expanding sleeve is tapped lightly into the hole until the sleeve is 20 mm below the surface (Figure 4.15). Verticality of bolt alignment relative to the surface can be checked

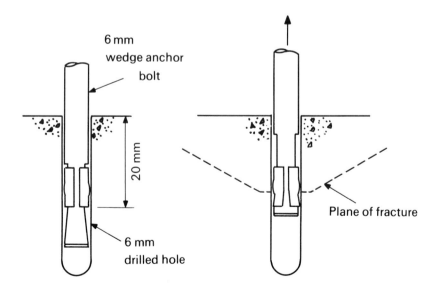

Figure 4.15 Internal fracture test.

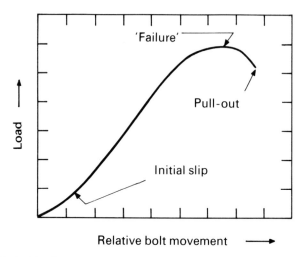

Figure 4.16 Typical loading curve.

using a simple slotted template. BS 1881: Part 207 (75) requires a minimum centre-to-centre spacing of 150 mm, and 75 mm edge distance.

The bolt is loaded at a standardized rate against a tripod reaction ring of 80 mm diameter with three feet, each 5 mm wide and 25 mm long. If necessary, shims may be used to correct for minor bolt misalignments. After applying an initial load to cause the sleeve to expand, the force required to produce failure by internal fracture of the concrete is measured. This will be the peak load indicated by the typical load/movement pattern in Figure 4.16. If the load is reduced once this peak has been reached there is likely to be no visible surface damage and it has been suggested that the bolt can be sawn off. If load application continues, a cone of concrete will be pulled from the surface, often intact, and considerable making good may be necessary. It has been found by the authors (97) that the load application method greatly influences the value of pull-out force required. The rate of load application affects not only the magnitude but also the variability of the results, and continuous methods yield more consistent results than if pauses are involved. Whatever loading method is adopted, it is essential that any calibration curves which are used relate specifically to the procedures followed. The importance of this is illustrated by comparison of the curves for two specific methods shown in Figure 4.20.

In the BRE loading method it is recommended that load is applied through a nut on the greased bolt thread by means of a torquemeter, which is rotated one half turn in 10 seconds and released before reading, the procedure being repeated until a peak is passed. The tripod assembly (Figure 4.17) incorporates a ball race and a facility for automatic alignment with the axis of the anchor bolt to ensure that an axial load is applied with no bending effects. Early

Figure 4.17 Torquemeter loading method.

tests also required a load cell, but subsequently the method was devloped on the basis of calibrations between measured torque and compressive strength.

Although this loading method is simple to use on both horizontal and vertical surfaces it suffers from two main disadvantages. Firstly, some torque is inevitably applied to the bolt, depending to some extent on the amount of grease on the thread, and this may reduce the failure load and increase the scatter obtained from individual results. Secondly, the torquemeter is relatively insensitive, and determination of the peak load is hindered by the use of settling pauses in the loading procedure.

An alternative mechanical loading method has been developed by the authors (97) which has the advantage of providing a direct pull free of twisting action. The latest version of this equipment is shown in Figure 4.18. The use of a proving ring for load measurement is sensitive, and provides a continuous rather than a settled reading, with the result that the variabilities due to load application and measurement are reduced. Loading is provided at a steady rate, without pauses, by rotating the loading handle at the rate of one revolution every 20 seconds. Calibration charts have been produced for this loading procedure which relate compressive strength to direct force, and the

Figure 4.18 Proving ring loading method.

variability due to testing using this approach has been shown to be lower than for the BRE method.

The load transfer mechanisms in these methods are complex, due to the concentrated localized actions of the expanding sleeve. The location of large aggregate particles relative to the sleeve will further complicate matters and affect the distribution of internal stresses. This is partially responsible for the high test variability found for the internal fracture test. The basic test-assembly dimensions have been determined largely from practical considerations of suitable magnitude of force, and obtaining a depth of test generally to avoid surface carbonation effects while minimizing likely reinforcement interference. As the name of the method implies, failure is thought to be initiated by internal cracking. Attempts have been made to represent this theoretically on the basis of an observed average failure depth of 17 mm which corresponds to a failure half angle of 78°. This is considerably greater than the likely angle of friction for the concrete of 37°, and application of the modified Coulomb failure criterion (as for the Lok-test, section 4.2.1.1) indicates failure by a combination of sliding and separation. This confirms the dependence of the pull-out force upon the tensile strength of the concrete, but in practice the test method at its present stage of development relies upon empirical calibrations.

Tests on cubes which were subsequently crushed have been described for a variety of mixes by both Chabowski (96) and the authors (97). Both reports indicate a reduction in crushing strength of 150 mm cubes of the order of 5% as a result of previous internal fracture tests on the cube. This must be taken into account when developing a calibration, unless undamaged specimens are available for comparison. There is also agreement that for practical purposes, mix characteristics (cement type, aggregate type, size and proportions) will not affect the pull-out/compressive strength relationship for natural aggregates. An upper limit of 20 mm on maximum aggregate size is suggested in view of the small test depth. The authors have also shown that the variability of results increases with aggregate size, and that moisture condition and maturity have negligible effects. These features represent the chief advantage of this approach compared with other non-destructive or partially-destructive methods, although the scatter of results is high, as illustrated by Figure 4.19 which represents the averages of six tests on a cube.

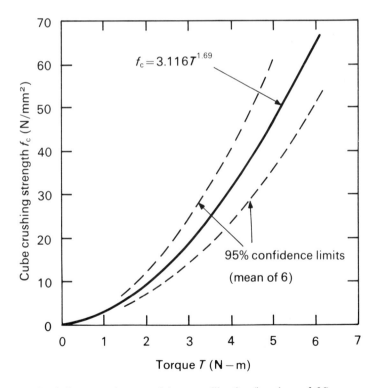

Figure 4.19 Typical compressive strength/torque calibration (based on ref. 96).

This means that a considerable number of specimens is required to produce a calibration curve.

The effects of precompression, as may be experienced in columns or prestressed construction, have also been examined by Chabowski (96) who concludes that there is no clearly defined influence. Although a trend towards an increase in pull-out force with increasing lateral compression is indicated, it is suggested that (provided zones of low stress are selected) this effect can be ignored in practice. The authors (97) have reported tests on beams in flexure which demonstrate a similar conclusion, although the variability of results appears to increase with increasing lateral bending compressive stress. The presence of direct lateral tensile stress will have a similar effect, and tests must not be made adjacent to visible cracks. Surface carbonation is another effect which both investigators conclude can be neglected in most circumstances. Only in very old concrete where the depth of carbonation approaches the depth of the test will this effect have any influence. The shallowness of test also offers the advantage that reinforcement is unlikely to affect results, but BS 1881: Part 207 (75) requires that it must be at least one bar diameter, or maximum aggregate size, outside the expected conic fracture surface.

The influence of the loading method has been indicated above. Results for the torquemeter loading method are generally expressed in the form of a compressive strength/torque relationship, but an average force/torque ratio of 1.15 is reported by Chabowski (96). Comparative tests by the authors (97) between the direct pull and torquemeter methods suggest a corresponding ratio of 1.4 which reflects the differences in loading technique. The average relationships for the techniques are compared in Figure 4.20. It must also be pointed out that a calibration obtained by the author using the torquemeter suggests compressive strengths up to 20% lower than the BRE calibration. A similar feature has also been indicated by Keiller (100) and Long (101), and cannot be ignored.

The variability of test results is high for a variety of reasons. These include the localized nature of the test, the imprecise load transfer mechanisms and variations due to drilling. 95% confidence limits on estimated strength of $\pm 30\%$ based on the mean of six test results are accepted for the torquemeter load method (75), provided that individual results causing a coefficient of variation of greater than 16% are discarded. The authors (97) have claimed a corresponding range of $\pm 20\%$ based on four results for 10 mm aggregates using the direct pull equipment. The average coefficients of variation observed for cubes of 20 mm maximum aggregate size were 16.5% for torque and 7.0% for direct force, with values 20% lower for 10 mm aggregate.

The method can be applied to lightweight concretes (24) although difficulties may be encountered with very soft aggregate types. Typical correlations using the torquemeter method are compared in Figure 4.21 from which the effects of aggregate type can be clearly seen. The measured torque corresponding to a given compressive strength is reduced in comparison to

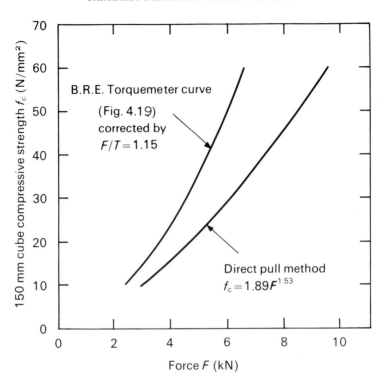

Figure 4.20 Comparison of calibration curves for natural aggregates (based on refs 96 and 97).

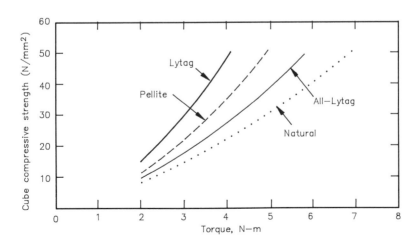

Figure 4.21 Comparison of calibration curves for lightweight aggregates (torquemeter method) (based on ref. 24).

natural aggregates, but is also significantly affected by the type of lightweight aggregate present. It is also interesting to note that a direct-pull load method gave much closer agreement between correlations for different aggregate types (24).

The chief advantage of the internal fracture test lies in the ability to use a general strength calibration curve for natural aggregates relating only to the loading method. Despite the variability, localized surface nature, and damage caused, this may be of particular value in situations where a strength estimate of in-situ concrete of unknown age or composition is required. This is especially true for slender members with only one exposed surface where cores or other direct techniques are not possible. The accuracy of strength estimate will be similar to that obtained by small cores but with considerable savings of time, expense and disruption.

4.2.2.2   ESCOT.   This test has been developed by Domone and Castro (99) to provide a cheap, robust and simple strength measurement technique. Designed as a drilled-hole method, the test works on an expanding sleeve principle which causes internal fracture of the concrete at a depth of just under 20 mm below the surface, with a conical failure zone of between 100 and 200 mm diameter. The test is much simpler than the Capo test, and although load is applied by a torquemeter, no bearing ring is required as in the internal fracture test. The equipment and principle are illustrated in Figure 4.22.

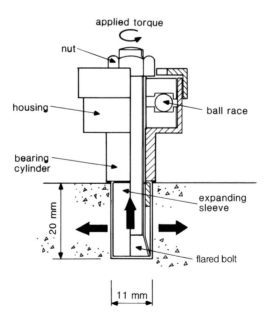

Figure 4.22 ESCOT expanding sleeve test system (based on ref. 99).

Tests to illustrate the effects of hole size (which should not exceed 0.2 mm oversize), depth and inclination (which should not exceed 10° from perpendicular) are reported (99) as well as the effects of reinforcement within the failure zone. The test can also be used in a hole preformed by a steel dowel during casting, but in this case measured values will be slightly reduced. Laboratory correlations with compressive strength are shown to be similar in nature but better than for the BRE internal fracture test, and comparable in accuracy to a modified ASTM pull-out approach described in section 4.2.1.2. On the basis of the average of four results, it is suggested that strength can be estimated to $\pm 25\%$ with no prior knowledge of the concrete other than aggregate size and type. Increased accuracy is likely with calibration for a particular concrete, and the possible benefits from combination with ultrasonic pulse velocity readings are suggested.

Further investigation is clearly required before the test can be generally accepted, but it does appear to offer a simpler and more reliable alternative to the BRE internal fracture method when aggregate details are known.

4.2.2.3 The Capo test. This has been developed in Denmark (1) as an equivalent to the Lok-test for situations where use cannot be preplanned. The basic geometry of the Lok-test described in section 4.2.1.1 has been maintained, although the pull-out insert consists of an expanding ring inserted into an undercut groove. The name is based on the expression 'cut and pull-out', and the procedure consists of drilling a 45 mm deep, 18 mm diameter hole, after which a 25 mm groove is cut at a depth of 25 mm using a portable milling machine illustrated in Figure 4.23. The expanding ring insert is then placed and expanded in the groove, as shown in Figure 4.24, and conventional Lok-test pulling equipment can be used as described previously. Testing must continue to pull out the plug of concrete, and the ring may be recovered, recompressed and re-used up to three or four times.

Extensive laboratory testing prgrammes (86) have shown that the behaviour of this test is effectively identical to the Lok-test and that the strength calibration and reliability may be regarded as the same. It is claimed that the entire testing operation, including drilling, may be completed in about ten minutes, and the equipment is available in the form of a comprehensive kit. In Denmark this method has been accepted as equivalent to the Lok-test and has been used on a number of projects for in-situ strength determination in critical zones (1). The potential areas of application are wide and although surface zone effects must be considered, the approach appears to offer the most reliable available indication of in-situ strength apart from cores. Although equipment costs are high, the damage, time and cost of testing will be considerably less than for cores. Problems may arise from the presence of reinforcement within the test zone, and bars must be avoided, but the value of this test is considerable in situations where mix details are not known.

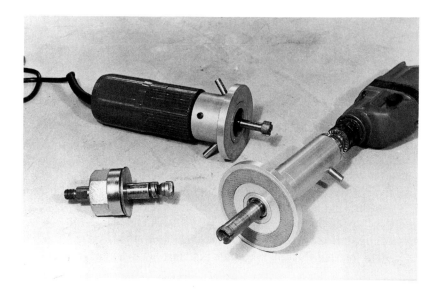

Figure 4.23 Capo test equipment.

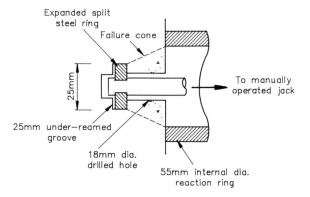

Figure 4.24 Capo test configuration.

4.2.2.4 Wood-screw method. A simple pull-out technique utilizing wood-screws has been described by Jaegermann (102). This is intended for use to monitor strength development in the 5–15 N/mm² strength range for formwork stripping purposes in industrialized buildings.

A nail is driven into the surface of the fresh concrete to push aside aggregate particles, and the screw with a plastic stabilizing ring attached at the appropriate height is inserted until the ring touches the concrete surface. The unthreaded upper part of of the screw is painted to prevent bonding and tests can be made at different depths by using screws of different lengths. Load is applied to the screw head by means of a proving ring or hydraulic jacking device. The principal assumption is that the force required to pull the screw from the concrete is dominated by the fine mortar surrounding the screw threads, and laboratory trials suggest good strength correlations with reasonable repeatability but further development is needed to facilitate field usage.

4.3   Pull-off methods

This approach has been developed to measure the in-situ tensile strength of concrete by applying a direct tensile force. The method may also be useful for measuring bonding of surface repairs (103) and a wide selection of equipment is commercially available (104) with disk diameters typically 50 mm or 75 mm. Procedures are covered by BS 1881: Part 207 and it should be noted that the fracture surface will be below the concrete surface and will thus leave some surface damage that must be made good.

Pull-off tests have been described (101) which were developed initially in the early 1970s for suspect high alumina concrete beams. A disk is glued to the concrete surface with an epoxy resin and jacked off to measure the force necessary to pull a piece of concrete away from the surface. The direct tension failure is illustrated in Figure 4.25, and if surface carbonation or skin effects are present these can be avoided by the use of partial coring to an appropriate depth. 'Limpet' loading equipment with a 10 kN capacity is commercially available to apply a tensile force through a rod screwed axially into a 50 mm

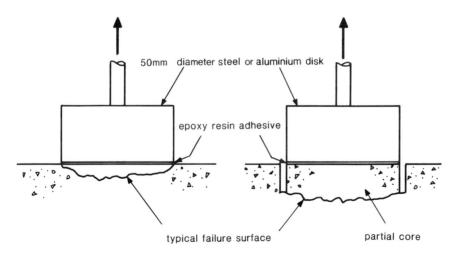

Figure 4.25 Pull-off method — surface and partially cored.

Figure 4.26 'Limpet' equipment.

diameter disk. This equipment (Figure 4.26) bears on the concrete surface adjacent to the test zone and is operated manually by steady turning of the handle, with the load presented digitally. Another common type of loading system is by means of a tripod apparatus, with the load applied mechanically (as in Figure 4.27) or hydraulically. Despite wide variations in loading rates and reaction configurations between different systems the authors (105) have concluded that results are unlikely to be affected provided there is adequate clearance between the disk and reaction points. Considerable care is needed in surface preparation of the concrete by sanding and degreasing to ensure good bonding of the adhesive, which may need curing for between 1.5 and 24 hours according to material and circumstances. Difficulties may possibly be encountered with damp surfaces.

BS 1881: Part 207 requires that the mean of six valid tests should be used, and that these should be centred at least two disk diameters apart. The stiffness of the disk has been shown to be an important parameter and the limiting thickness/diameter ratio will depend upon the material used (105). This is illustrated in Figure 4.28 from which it can be seen that to ensure a uniform stress distribution, and hence maximum failure load, steel disk

Figure 4.27 'Hydrajaws' tripod equipment.

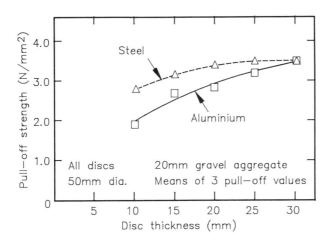

Figure 4.28 Effects of disk type and thickness (based on ref. 105).

thickness must be 40% of the diameter whilst for aluminium this rises to 60%. These experimental findings have been supported by finite element analyses.

A nominal tensile strength for the concrete is calculated on the basis of the disk diameter, and this may be converted to compressive strength using a calibration chart appropriate to the concrete. This calibration will differ according to whether coring is used or not (100), with cored tests generally requiring a lower pull-off force. Partial coring will transfer the failure surface lower into the body of the concrete, but the depth of coring may also be critical, as illustrated by Figure 4.29, and should always exceed 20 mm. Reinforcing steel clearly must be avoided when partial coring is used. A test coefficient of variation of 7.9% with a range of predicted/actual strength between 0.85 and 1.25 related to 150 mm Portland cement cubes has been reported by Long and Murray (101) using the mean of three test results. A typical calibration curve is illustrated in Figure 4.30, and it is claimed that factors such as age, aggregate type and size, air entrainment, compressive stress and curing have only marginal influences upon this. Extensive field tests during the construction of a multistorey car park have also been successfully undertaken (106).

BS 1881: Part 207 recommends that a strength correlation should be established for the concrete under investigation and that site results from one location are likely to yield a coefficient of variation of about 10%. Accuracies of strength predictions under laboratory conitions of about ±15% (95% confidence limits) are likely. The authors (105) have also demonstrated that separate correlations are required for different types of lightweight aggregates, as illustrated in Figure 4.31, and that these are different to those for natural aggregates due to different tensile/compressive strength relationships. It can be noted that pull-off values for lightweight aggregates are higher than for natural aggregates at a given strength level.

This test is aimed primarily at unplanned in-situ strength determination. The method is particularly suitable for small-section members, and long-

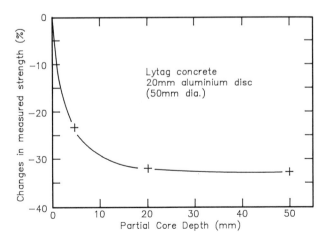

Figure 4.29 Effects of partial coring (based on ref. 105).

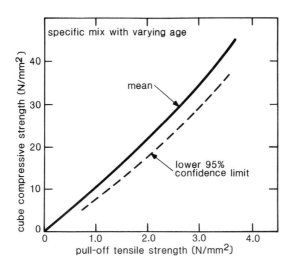

Figure 4.30 Typical pull-off/strength correlation for natural aggregate (based on ref. 106).

term monitoring procedures could also be developed involving proof load tests at intervals on a series of permanent probes. It is also particularly suited, with the use of partial coring into the base material, for assessment of bonding strength of repairs as indicated above.

This is an area receiving considerable industrial interest and many repair specifications now require pull-off testing as part of quality control procedures (107). In such cases it is usual to specify a minimum pull-off stress and it is thus vital that the test procedures are carefully specified or standardized if this is to be meaningful.

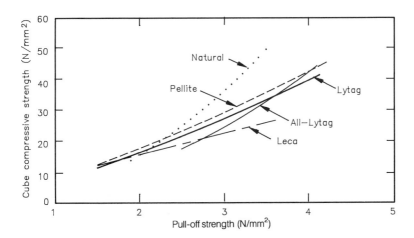

Figure 4.31 Typical strength correlations for lightweight aggregates.

4.4 Break-off methods

4.4.1 *Norwegian method*

Johansen (108) has reported the use of a break-off technique developed in Norway. This is intended primarily as a quality control test, and makes a direct determination of flexural strength in a plane parallel and at a certain distance from the concrete surface. A tubular disposable form is inserted into the fresh concrete, or alternatively a shaped hole can be drilled, to form a slot of the type shown in Figure 4.32. The core left after the removal of the insert is broken off by a transverse force applied at the top surface as shown. This force is provided hydraulically using specially developed portable equipment available under the name 'TNS-Tester'.

The 'break-off' strength calculated from the results has been shown to give a linear correlation with the modulus of rupture measured on prism specimens, although values were 30% higher on average. Christiansen *et al.* (109) have also examined relationships between break-off values and bending tensile strengths and have shown water/cement ratio, age, curing and cement type to be significant. Values obtained (108) for an airfield pavement contract have suggested coefficients of variation of 6.4% for laboratory samples and 12.6% in-situ. Comparable values have also been found on other construction

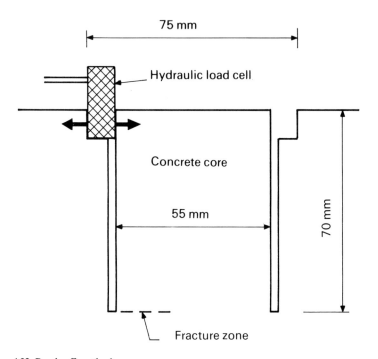

Figure 4.32 Break-off method.

sites (110). It is suggested that the mean of five test results should be used in view of high within-test variation. BS 1881: Part 207 (75) requires a concrete element thickness of at least 100 mm, with a minimum clear spacing or edge distance from the outer face of the groove of four times the maximum aggregate size ($\not< 50$ mm). Reinforcing bars must obviously be avoided and particular care is needed to ensure that compaction and curing at prepared test positions are representative of the surrounding body of concrete.

It is claimed that the method is quick and uncomplicated, taking less than two minutes per test. Results are not significantly affected by the surface condition or local shrinkage and temperature effects. A correlation with compressive strength has been developed which covers a wide range of concrete, but this is likely to be less reliable than a tensile strength correlation in view of the factors influencing the tensile/compressive strength relationship. Compressive strength estimates to within $\pm 20\%$ should be possible with the aid of appropriate calibrations. The method is regarded as especially suitable for very young concrete, and although leaving a sizeable damage zone, may gain acceptance as an in-situ quality control test where tensile strength is important. Although quicker than compression testing of cores, the use of results for strength estimation of old concrete may be unreliable unless a specific calibration relationship is available. Field experience in a variety of situations has been reported by Carlsson *et al.* (111). Naik (112) has also reviewed field experience and indicated that crushed aggregates may give strengths about 10% higher than those of rounded aggregates, and that drilled tests give results 9% higher than when a sleeve is inserted into the fresh concrete.

Dahl-Jorgensen and Johansen (110) also describe a modified version of the test in which a steel cylinder is glued to the surface and jacked against a counter-support. This is intended for bond testing of concrete overlays or epoxy surface coatings, and lower test variability is claimed than with pull-off methods.

4.4.2 *Stoll tork test*

This proposed approach (113) is intended to improve upon the variability encountered with other methods and to permit tests at greater depths below the surface than pull-out, pull-off or penetration resistance methods. A cylindrical cleated spindle of 18 mm height and 35 mm diameter (Figure 4.33) is removably attached to a 19 mm torque bolt at least 51 mm in length and cast into the concrete at the required depth. A compressible tape is attached to the periphery except for the radially extending symmetrically opposed cleat surfaces which bear directly against the concrete mortar. A small grating prevents intrusion of large aggregate particles into the mortar cusp which is fractured by application of torque to the bolt by a conventional torque wrench.

The maximum load is correlated to the compressive strength. The mortar cusp is subjected to a semi-confined compressive stress leading to a

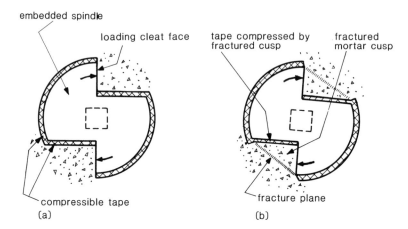

Figure 4.33 Stoll tork test (based on ref. 113). (a) Before fracture. (b) After fracture.

compressive/shear failure. Limited data available at present indicate reliable linear relationships with cylinder strength in the range 6.7–34 N/mm², but affected by aggregate type, cement replacements and admixtures. Variability and accuracy compare favourably with, but are not significantly better than, other methods discussed in this chapter, and results should be based on the average of at least three tests. The principal value of the method appears to lie in preplanned monitoring of internal in-situ strength development, perhaps related to formwork stripping or load application. It is understood that equipment is being refined for commercial production, and further investigation of factors influencing strength correlations is clearly required.

4.4.3 Shearing-rib method

This is a long established test for precast concrete quality control purposes in the former Soviet Union, which has been described by Leshchinsky *et al.* (34). A hand-operated hydraulic jack is clamped to a linear element which is at least 170 mm thick and is used to shear-off a corner of the element. The localized load is applied over a 30 mm width at an angle of 18° to the surface and at a distance of 20 mm from the edge. Reinforcing steel set at normal covers will thus not influence results and a highly stable strength correlation is claimed. At present, this technique has not become established elsewhere in the world.

5 Cores

The examination and compression testing of cores cut from hardened concrete is a well-established method, enabling visual inspection of the interior regions of a member to be coupled with strength estimation. Other physical properties which can be measured include density, water absorption, indirect tensile strength and movement characteristics including expansion due to alkali-aggregate reactions. Cores are also frequently used as samples for chemical analysis following strength testing. In most countries standards are available which recommend procedures for cutting, testing and intepretation of results; BS 1881: Part 120 (114) in the UK, whilst ASTM C42 (115) and ACI 318 (116) are used in the USA. Extremely valuable and detailed supplementary information and guidance is also given by Concrete Society Technical Report 11 (25) and its addendum.

5.1 General procedures for core cutting and testing

5.1.1 *Core location and size*
Core location will be governed primarily by the basic purpose of the testing, bearing in mind the likely strength distributions within the member, discussed in Chapter 1, related to the expected stress distributions. Where serviceability assessment is the principal aim, tests should normally be taken at points where likely minimum strength and maximum stress coincide, for example from the top surface at near midspan for simple beams and slabs, or from any face near the top of lifts for columns or walls. If the member is slender, however, and core cutting may impair future performance, cores should be taken at the nearest non-critical locations. Aesthetic considerations concerning the appearance after coring may also sometimes influence the choice of locations. Alternatively, areas of suspect concrete may have been located by other methods.

If specification compliance determination is the principal aim, the cores should be located to avoid unrepresentative concrete, and for columns, walls or deep beams will normally be taken horizontally at least 300 mm below the top of the lift. If it is necessary to drill vertically downwards, as in slabs, the core must be sufficiently long to pass through unrepresentative concrete which may occupy the top 20% of the thickness. In such cases drilling upwards from the soffit, if this is feasible, may considerably reduce the extent of drilling, but the operation may be more difficult and may introduce additional uncertainties relating to the effects of possible tensile cracking. Reinforcement bars passing through a core will increase the uncertainty of strength testing, and should be avoided wherever possible. The use of a covermeter to locate reinforcement prior to cutting is therefore recommended.

Where the core is to be used for compression testing, British and American Standards require that the diameter is at least three times the nominal maximum aggregate size. BS 1881: Part 120 (114) further requires that a minimum diameter of 100 mm is used, with 150 mm preferred, although in Australia 75 mm is considered to be generally acceptable. In general, the accuracy decreases as the ratio of aggregate size to core diameter increases and 100 mm diameter cores should not be used if the maximum aggregate size exceeds 25 mm, and this should preferably be less than 20 mm for 75 mm cores. In some circumstances smaller diameters are used, especially in small-sized members where large holes would be unacceptable, but the interpretation of results for small cores becomes more complex and is considered separately in section 5.3. The choice of core diameter will also be influenced by the length of specimen which is possible. It is generally accepted that cores for compression testing should have a length/diameter ratio of between 1.0 and 2.0, but opinions vary concerning the optimum value.

The Concrete Society (25) and BS 1881: Part 120 (114) suggest that cores should be kept as short as possible ($l/d = 1.0 \rightarrow 1.2$) for reasons of drilling costs, damage, variability along length, and geometric influences on testing. Although these points are valid, procedures for relating core strength to cylinder or cube strength usually involve correction to an equivalent standard cylinder with $l/d = 2.0$, and it can be argued that uncertainties of correction factors are minimized if the core length/diameter ratio is close to 2.0 (117) (see section 5.2.2) and this view is supported by ASTM C42 (115).

The number of cores required will depend upon the reasons for testing the volume of concrete involved. The likely accuracies of estimated strength are discussed in section 5.2.3, but the numbers of cores must be sufficient to be representative of the concrete under examination as well as provide a strength estimate of acceptable accuracy as discussed in Chapter 1. ACI 318 (116) requires that at least three cores are always used.

5.1.2 Drilling

A core is usually cut by means of a rotary cutting tool with diamond bits, as shown in Figure 5.1. The equipment is portable, but it is heavy and must be firmly supported and braced against the concrete to prevent relative movement which will result in a distorted or broken core, and a water supply is also necessary to lubricate the cutter. Vacuum assisted equipment can be used to obtain a firm attachment for the drilling rig without resorting to expansion bolts or cumbersome bracing. Uniformity of pressure is important, so it is essential that drilling is performed by a skilled operator. Hand-held equipment is available for cores up to 75 mm diameter. A cylindrical specimen

Figure 5.1 Core cutting drill.

is obtained, which may contain embedded reinforcement, and which will usually be removed by breaking off by insertion of a cold chisel down the side of the core, once a sufficient depth has been drilled. The core, which will have a rough inner end, may then be removed using the drill or tongs, and the hole made good. This is best achieved either by ramming a dry, low shrinkage concrete into the hole, or by wedging a cast cylinder of suitable size into the hole with cement grout or epoxy resin. It is important that each core is examined at this stage, since if there is insufficient length for testing, or excessive reinforcement or voids, extra cores must be drilled from adjacent locations. Each core must be clearly labelled for identification, with the drilled surface shown, and cross-referenced to a simple sketch of the element drilled. Photographs of cores are valuable for future reference, especially as confirmation

Figure 5.2 Typical core.

of features noted during visual inspection, and these should be taken as soon as possible after cutting. A typical photograph of this type is shown in Figure 5.2.

5.1.3   Testing

Each core must be trimmed and the ends either ground or capped before visual examination, assessment of voidage, and density determinations.

5.1.3.1   Visual examination.   Aggregate type, size and characteristics should be assessed together with grading. These are usually most easily seen on a wet surface, but for other features to be noted, such as aggregate distribution, honeycombing, cracks, defects and drilling damage, a dry surface is preferable. Precise details of the location and size of reinforcement passing through the core must also be recorded. The voids should be classified in terms of the excess voidage by comparison with 'standard' photographs of known voidage provided by Concrete Society Technical Report 11 (25) and BS 1881: Part 120 (114). These reference photographs are based on the assumption of a fully compacted 'potential' voidage of 0.5%. This estimated value of excess voidage will be required when attempting to calculate the potential strength (see section 5.2.2). If a more detailed description of the voids is required, this should refer to small voids (0.5–3 mm), medium voids (3–6 mm), and large voids (>6 mm) with the term 'honeycombing' being used if these are interconnected.

5.1.3.2 Trimming. Trimming, preferably with a masonry or water-lubricated diamond saw, should give a core of a suitable length with parallel ends which are normal to the axis of the core. If possible, reinforcement and unrepresentative concrete should be removed.

5.1.3.3 Capping. Unless their ends are prepared by grinding, cores should be capped with high alumina cement mortar or sulphur–sand mixture to provide parallel end surfaces normal to the axis of the core. (Other materials should not be used as they have been shown to give unreliable results.) Caps should be kept as thin as possible, but if the core is hand trimmed they may be up to about the maximum aggregate size at the thickest points.

5.1.3.4 Density determination. This is recommended in all cases, and is best measured by the following procedure (25):

(i) Measure volume (V_u) of trimmed core by water displacement
(ii) Establish density of capping materials (D_c)
(iii) Before compressive testing, weigh soaked/surface-dry capped core in air and water to determine gross weight W_t and volume V_t
(iv) If reinforcement is present this should be removed from the concrete after compression testing, and the weight W_s and volume V_s determined
(v) Calculate saturated density of concrete in the uncapped core from

$$D_a = \frac{W_t - D_c(V_t - V_u) - W_s}{V_u - V_s}.$$

If no steel is present, W_s and V_s are both zero.

The value thus obtained may be used, if required, to assess the excess voidage of the concrete using the relationship

$$\text{estimated excess voidage} = \frac{D_p - D_a}{D_p - 500} \times 100\%$$

where D_p = the potential density based on available values for 28-days-old cubes of the same mix.

5.1.3.5 Compression testing. The standard procedure in the United Kingdom is to test cores in a saturated condition, although in the USA (116) dry testing is used if the in-situ concrete is in a dry state. If the core is to be saturated, testing should be not less than two days after capping and immersion in water. The mean diameter must be measured to the nearest 1 mm by caliper, with measurements on two axes at quarter- and mid-points along the length of the core, and the core length also measured to the nearest 1 mm.

Compression testing will be carried out at a rate within the range $12–24\,N/(mm^2 \cdot min)$ in a suitable testing machine and the mode of failure noted. If there is cracking of the caps, or separation of cap and core, the result should be considered as being of doubtful accuracy. Ideally cracking should be similar all round the circumference of the core, but a diagonal shear crack is considered satisfactory, except in short cores or where reinforcement or honeycombing is present.

5.1.3.6 Other strength tests on cores. Although compression testing as described above is by far the most common method of testing cores for strength, recent research has indicated the potential of other methods which are outlined below. Two of these measure the tensile strength, although neither method is yet fully established. Tensile strength may also be measured by 'Brazilian' splitting tests on cores according to ASTM C42 (115). Tests for other properties of the concrete, such as permeability, alkali/aggregate expansion, or air content (Chapters 7, 8 and 9) may also be performed on suitably prepared specimens obtained from cores.

In the point load strength test, Robins (118) has shown that the point load

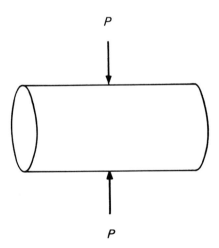

Figure 5.3 Point load test.

test, which is an accepted method of rock strength classification, may usefully be applied to concrete cores. A compressive load is applied across the diameter (Figure 5.3) by means of a manually operated hydraulic jack, with the specimen held between spherically truncated conical platens with a point of 5 mm radius. It has been found that the point load strength index is indirectly related to the concrete compressive strength, although core size and aggregate type affect the relationship. For a given aggregate and core size, the index varies linearly with cube strength for strengths greater than 20 N/mm². Robins (118) also claims that the testing variability is comparable to that expected for conventional core testing. The advantages of this approach are that trimming and capping are not required and that the testing forces are lower, thus permitting the use of small portable equipment on site at a reduced unit cost.

The point load test is essentially a tensile test, and Robins (119) has also confirmed that a simple linear relationship exists between point load index and flexural strength. This test may thus be particularly useful for sprayed and fibrous concretes (120).

In the gas pressure tension test, Clayton (121) has demonstrated that applied gas pressure may be used on cylinders to simulate the effects of uniaxial tensile tests, and that cores may also be used for this purpose. The specimen is inserted into a cylindrical steel jacket with seals at each end, and gas pressure is applied to the bare curved surface. Nitrogen has been found to be safe and convenient. The flow is controlled by a single-stage regulator, and a pressure gauge is used for measurement. Pressure is increased manually at a specified rate, until failure occurs by the formation of a single cleavage plane transverse to the axis of the specimen. The two sections are forced violently apart and safety precautions are necessary to prevent ejection of the fragments from the testing jacket.

The method has been developed using 100-mm cylinders, but has been successfully applied to 75-mm cores of high alumina cement concrete which in some cases had length/diameter ratios of less than 1.0. Although preliminary evidence suggests that this may provide a reliable method of determining in-situ tensile strength, further research is necessary before results can be regarded with confidence.

A third type of 'strength' test on cores which has been developed by Chrisp et al. (122) utilizes low strain rate compressive load cycling on 72 mm diameter cores with a length/diameter ratio of 2.5 to quantify damage in cases where deterioration has occurred. Strain data are recorded using a sensitive 'compressometer', and processed automatically on a microcomputer to yield hysteresis and stiffness characteristics. These can be used to establish a series of parameters of damage caused by deterioration, and cores are not significantly further damaged by the test, thus allowing them to be subjected to further testing. Good results have been achieved with this 'stiffness damage' test applied to concrete affected by alkali-aggregate reactions and the approach is likely to be extended to other damage mechanisms.

5.2   Interpretation of results

5.2.1   *Factors influencing measured core compressive strength*
These may be divided into two basic categories according to whether they
are related to concrete characteristics or testing variables.

5.2.1.1   Concrete characteristics.   The moisture condition of the core will
influence the measured strength — a saturated specimen has a value 10–15%
lower than a comparable dry specimen. It is thus very important that the
relative moisture conditions of core and in-situ concrete are taken into
account in determining actual in-situ concrete strengths. If the core is tested
while saturated, comparison with standard control specimens which are also
tested saturated will be more straightforward but there is evidence (123) that
moisture gradients within a core specimen will also tend to influence measured
strength. This introduces additional uncertainties when procedures involving
only a few days of either soaking or air drying are used since the effects of
this conditioning are likely to penetrate only a small distance below the surface.

The curing regime, and hence strength development, of a core and of the
parent concrete will be different from the time of cutting. This effect is very
difficult to assess, and in mature concrete may be ignored, but for concrete
of less than 28 days should be considered.

Voids in the core will reduce the measured strength, and this effect can be
allowed for by measurement of the excess voidage when comparing core
results with standard control specimens from the point of view of material
specification compliance. Figure 5.4, based on reference (25), shows the
influence of this effect. Under normal circumstances an excess voidage of less
than 2.5% would be expected.

5.2.1.2   Testing variables. These are numerous, and in many cases will have
a significant influence upon measured strength. The most significant factors
are outlined below.

(i) *Length/diameter ratio of core.* As the ratio increases, the measured
strength will decrease due to the effect of specimen shape on stress
distributions whilst under test. Since the standard cylinder used in many
parts of the world has a length/diameter ratio of 2.0, this is normally regarded
as the datum for computation of results, and the relationship between this
and a standard cube is established. Monday and Dhir (117) have indicated
the influence of strength on length/diameter effects and this is confirmed by
Bartlett and MacGregor (124) who also indicate the influence of moisture
conditions. It is claimed that correction factors to an equivalent length/diameter
ratio of 2.0 will move towards 1.0 for soaked cores and as concrete strength
increases. The authors have also demonstrated the influence of aggregate
type when lightweight aggregates are present (24). This issue is widely

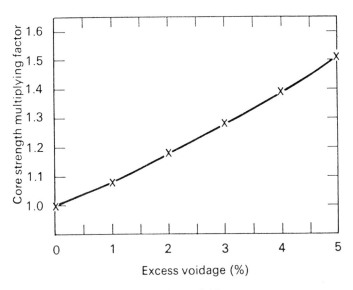

Figure 5.4 Excess voidage corrections (based on ref. 25).

recognized to be subject to many uncertainties, but the average values shown in Figure 5.5, based on the Concrete Society recommendations (25), have been adopted by BS 1881. These differ from ASTM (115) suggestions which recognize but do not allow for, strength effects and are also limited to cylinder strengths in the range 13–41 N/mm².

(ii) *Diameter of core.* The diameter of core may influence the measured strength and variability (see section 5.1.1). Measured concrete strength will generally decrease as the specimen size increases; for sizes above 100 mm this effect will be small, but for smaller sizes this effect may become significant. As the diameter decreases, the ratio of cut surface area to volume increases, and hence the possibility of strength reduction due to cutting damage will increase. It is generally accepted that a minimum diameter/maximum aggregate size ratio of 3 is required to make test variability acceptable.

(iii) *Direction of drilling.* As a result of layering effects, the measured strength of a specimen drilled vertically relative to the direction of casting is likely to be greater than that for a horizontally drilled specimen from the same concrete. Published data on this effect are variable, but an average difference of 8% is suggested (25) although there is evidence that this effect is not found with lightweight aggregate concretes (24). Whereas standard cylinders are tested vertically, cubes will normally be tested at right angles to the plane of casting and hence can be related directly to horizontally drilled cores.

(iv) *Method of capping.* Provided that the materials recommended in section 5.1.3.3 have been used, their strength is greater than that of the core, and the caps are sound, flat, perpendicular to the axis of the core and not excessively thick, the influence of capping will be of no practical significance.

(v) *Reinforcement.* Published research results indicate that the reduction in measured strength due to reinforcement may be less than 10%, but the variables of size, location and bond make it virtually impossible to allow for accurately. Reinforcement must therefore be avoided wherever possible, but in cases where it is present the measured core strength may be corrected but treated with caution. Recent developments in coring technology in Germany (14) have resulted in a drilling machine with an automatic detection and stop facility before reinforcement is cut. It is suggested (114) that for a core containing a bar perpendicular to the axis of the core the following correction factor may be applied to the measured core strength (but disregard core if correction > 10%):

$$\text{corrected strength} = \text{measured strength} \times \left[1.0 + 1.5\left(\frac{\phi_r}{\phi_c} \cdot \frac{h}{l}\right)\right]$$

where ϕ_r = bar diameter
      ϕ_c = core diameter
       h = distance of bar axis from nearer end of core
       l = core length (uncapped).

Multiple bars within a core can similarly be allowed for by the expression

$$\text{corrected strength} = \text{measured strength} \times \left[1.0 + 1.5\frac{\Sigma(\phi_r \cdot h)}{\phi_c \cdot l}\right]$$

If the spacing of two bars is less than the diameter of the larger bar, only the bar with the higher value of $(\phi_r \cdot h)$ should be considered.

5.2.2 Estimation of cube strength
Estimation of an equivalent cube strength corresponding to a particular core result must initially account for two main factors. These are

(i) The effect of the length/diameter ratio, which requires a correction factor, illustrated by Figure 5.5, to be applied to convert the core strength to an equivalent standard cylinder strength
(ii) Conversion to an equivalent cube strength using an appropriate relationship between the strength of cylinders and cubes.

Corrections for the length/diameter ratio of the core have been discussed in section 5.2.1.2. Subsequent conversion to a cube strength is usually based on the generally accepted average relationship that cube strength = 1.25 cylinder strength (for $l/d = 2.0$). Monday and Dhir (117) have shown that this

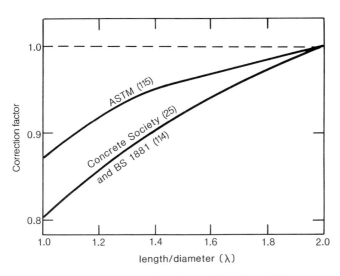

Figure 5.5 Length/diameter ratio influence (based on refs 25, 114 and 115).

relationship is a simplification, and that a more reliable conversion can be obtained from

$$\text{cube strength} = Af_{cy} - Bf_{cy}^2$$

where f_{cy} is the strength of a core with $l/d = 2.0$, and constants $A = 1.5$ and $B = 0.007$ tentatively.

Within the range of 20–50 N/mm² cylinder strengths this produces values within 10% of those given by the average factor 1.25, but the discrepancies increase for lower- and higher-strength concrete. Such discrepancies will, however, be partially offset for cores with l/d close to 1.0 by the errors resulting from the use of (l/d) ratio correction factors which are not strength-related. Particular care will be needed to take account of this issue when dealing with cores of high-strength concrete, which is increasingly being used worldwide.

The Concrete Society (25) recommends a procedure incorporating the correction factors of Figure 5.5, coupled with an allowance of 6% strength differential between a core with a cut surface relative to a cast cylinder. A strength reduction of 15% is also incorporated to allow for the weaker top surface zone of a corresponding cast cylinder, before conversion to an equivalent cube strength by the multiplication factor of 1.25. An 8% difference between vertical and horizontally drilled cores is also incorporated with the resulting expressions (as adopted by BS 1881) emerging.

Horizontally drilled core:

$$\text{estimated in-situ cube strength} = \frac{2.5f_\lambda}{1.5 + 1/\lambda}$$

Vertically drilled core:

$$\text{estimated in-situ cube strength} = \frac{2.3f_\lambda}{1.5 + 1/\lambda}$$

where f_λ is the measured strength of a core with length/diameter $= \lambda$.

It is interesting to note that, using these expressions, the strength of a horizontally drilled core of length/diameter $(\lambda) = 1$ will be the same as the estimated cube strength. The cube strengths evaluated in this way will be estimates of the actual in-situ strength of the concrete in a wet condition and may underestimate the strength of the dry concrete by 10–15%.

The strength differences between in-situ concrete and standard specimens have been fully discussed in Chapter 1. An average recommended relationship is that the 'potential' strength of a standard specimen made from a particular mix is about 30% higher than the actual 'fully compacted' in-situ strength (25). If this value is used to estimate a potential strength for comparison with specifications, the uncertainty of the relationship must be remembered. Appendix 3 of the Concrete Society Report offers detailed guidance relating to curing history, but potential strength estimations are increasingly unpopular due to the difficulties of accounting for all variable factors.

The expressions for cube strength will change as follows:

Horizontally drilled core:

$$\text{estimated potential cube strength} = \frac{3.25f_\lambda}{1.5 + 1/\lambda}.$$

Vertically drilled core:

$$\text{estimated potential cube strength} = \frac{3.0f_\lambda}{1.5 + 1/\lambda}.$$

A worked example of evaluation of core results using the Concrete Society recommendations is given in Appendix C of this book.

ACI 318 (116) suggests that an average in-situ strength of at least 85% the minimum specified value is adequate, and that cores may be tested after air-drying for 7 days if the structure is to be dry. This is based on equivalent cylinder strengths derived from ASTM C42 (115) factors.

The effect of the calculation method can be considerable, as illustrated in Figure 5.6, and this emphasizes the importance of agreement between all parties of the method to be used in advance of the testing.

5.2.3 Reliability, limitations and applications
The likely coefficient of variation due to testing is about 6% for carefully cut and tested cores, which can be compared with a corresponding value of 3% for cubes. The difference is largely caused by the effects of cutting,

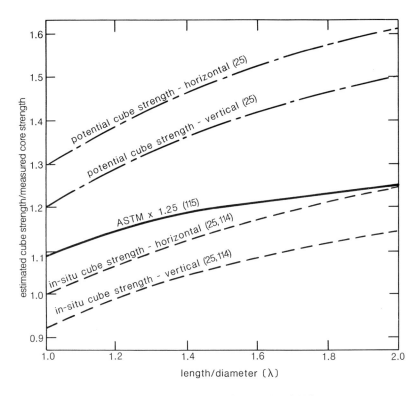

Figure 5.6 Effect of calculation method (baed on refs 25, 114 and 115).

especially since cut aggregate particles are only partially embedded in the core and may not make a full contribution during testing. It is claimed that the likely 95% confidence limits on actual strength prediction for a single core are $\pm 12\%$ when the Concrete Society calculation procedures (25) are adopted. It follows that for a group of n cores, the 95% confidence limits on estimated actual in-situ strengths are $\pm 12/\sqrt{n}\%$ (see also section 1.6.3). Where the 'potential' strength of the concrete is to be assessed, a minimum of four cores is required and an accuracy of better than $\pm 15\%$ cannot be expected. This can only be achieved if great care is taken to ensure that the concrete tested is representative, by careful location and preparation of the specimens. Uncertainties caused by reinforcement, compaction or curing may lead to an accuracy as low as $\pm 30\%$.

Examination of Figure 5.6 shows the differences between the results for in-situ and potential strength computed by the methods currently in use in the UK. Cube strengths derived from ASTM C42 (115) procedures coupled with an average cube/cylinder factor of 1.25 are also indicated and it will be clear that results computed in this way are liable to overestimate the actual strength by up to 16%. The Concrete Society method makes detailed allowance for the many variable factors influencing core results, and will provide the more reliable estimates of equivalent cube strengths.

Damage caused by drilling may be particularly significant for old brittle concretes, where internal cracking of the core may be aggravated by the loss of the confining effect of the surrounding body of concrete. Difficulties associated with core testing of concrete which has been damaged by alkali-aggregate reactions have similarly been identified (125). Tests on cores from flexurally cracked tensile zones must be regarded as unreliable, whilst Yip (126) has demonstrated that compressive load history of the concrete before coring may lead to reductions in measured strength of up to 30% due to internal microcracking. This latter effect may be quite significant even at relatively low stress levels and adds to interpretation uncertainties. The estimated cube strengths obtained from core compression tests may tend to underestimate the true in-situ capacity in all these situations. Strength changes with age may also be considered when interpreting core results, but any allowances must be carefully considered as discussed in section 1.5.2.

The principal limitations of core testing are those of cost, inconvenience and damage, and the localized nature of the results. It is strongly recommended that core testing is used in conjunction with some other form of testing which is less tedious and less destructive. The aim is to provide data on relative strengths within the body of the concrete under test. The size of core needed for reliable strength testing can pose a serious practical problem; 'small' cores may be worthy of consideration with slender members. It may also be appropriate to consider using a larger number of 'small'-diameter cores to obtain an improved spread of test locations where large volumes of concrete are involved. The cutting effort for three 50-mm cores may be as low as one-third of that for one 150 mm specimen. A comparable overall strength accuracy may be expected (see section 5.3.2) provided that maximum aggregate size is less than 17 mm. Where cores are used for other purposes, it will often be possible to use a 'small' diameter with considerable savings of cost, inconvenience and damage.

Apart from physical testing, cores often provide the simplest method of obtaining a sample of the in-situ concrete for a variety of purposes, but care must be taken that the effects of drilling, including heat generated by friction, or the presence of water, do not distort the subsequent results. A sample taken from the centre of a core may conveniently overcome this problem. Chemical analysis can often be performed on the remains of a crushed core, or specimens may be taken specifically for that purpose. Visual inspection of the interior of the concrete may be extremely valuable both for the assessment of compaction and workmanship, and for obtaining basic data about concrete for which no records are available. In cases where structural assessments of old structures are required, cores may also prove valuable in confirming covermeter results concerning the location and size of reinforcement.

5.3 Small cores

Although standards normally require cores to have a minimum diameter of 100 mm for compressive strength testing, cores of smaller diameter offer considerable advantages in terms of reduced cutting effort, time and damage. For applications such as visual inspection, density, or voidage determination, reinforcement location or chemical testing, these savings may be valuable. However, the reliability of small diameter cores for compression testing is lower than for 'normal' specimens. The many factors which affect normal core results may also be expected to influence small cores, but the extent of these factors may vary and other effects which are normally unimportant may become significant.

5.3.1 *Influence of specimen size*

It is well-established that measured concrete strength usually increases as the size of the test specimen decreases, and that results tend to be more variable. This latter effect has been shown to be particularly true for core specimens since the ratio of cut surface area to volume increases as diameter decreases and hence the potential influence of drilling damage is increased. Also, the ratio of aggregate size to core diameter is increased and may possibly exceed the generally recognized acceptable limit of 1:3. It is also well established that concrete strength is a further factor that may influence the behaviour of a core. These various factors are inter-related and difficult to isolate. For example, increased strength due to small specimen size may be offset by a reduction due to cutting effects.

The most common diameter for small cores is 40–50 mm. The authors have reported extensive laboratory tests to investigate the behaviour of 44 mm specimens (127), in which a total of 23 mixes were used, ranging from 10 to 82 N/mm^2 with 10 and 20 mm gravel aggregates, and cores were cut from 500 × 100 × 100 mm laboratory cast prism specimens to provide a range of length/diameter ratios.

5.3.1.1 Length/diameter ratio. The average relationship for length/diameter effects on 44 mm cores (127) is shown in Figure 5.7 which compares the relationships discussed in section 5.2.1.2 for normal cores. This relationship was found to be independent of drilling orientation, aggregate size and cement type, for practical purposes, although the scatter of results is high since each point in Figure 5.7 represents the average of four similar cores. It will be seen that the correction factor for length/diameter ratio is reasonably close to the Concrete Society recommendation (25) for larger cores. Lightweight concretes are likely to have values closer to 1.0 (24).

5.3.1.2 Variability of results. No significant change of variability was found between the extremes of length/diameter ratio for either aggregate size, and

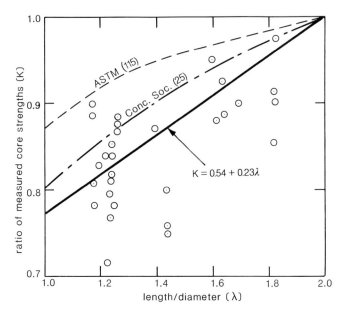

Figure 5.7 Length/diameter ratio for small cores (based on refs 25, 115 and 127).

the average coefficient of variation of 8% was also independent of orientation. However, taking account of concrete variability as indicated by control cubes, it is clear that 20 mm aggregate cores show a higher variability due to cutting and testing than 10 mm aggregates (127). The range of coefficients of variation of strength for groups of similar cores was large, and made identification of the effects of other variables impossible to assess. Bowman (128) has reported a coefficient of variation of 28.9% for 50 mm cores from in-situ concrete on a site in Hong Kong compared with a value of 19.5% for corresponding 150 mm cores from the same concrete. Swamy and Al-Hamed have also suggested that variability reduces as strength increases (129), whilst lightweight aggregate concretes may also be less variable (24).

5.3.1.3 Measured strength. Based on the authors' tests (127) the factors required to convert the measured core strength (after correction to $\lambda = 2.0$) to an equivalent 100 mm cube strength are given in Table 5.1. If an equivalent 150 mm cube strength is required, these values may be reduced by 4%.

It can be seen that for 10-mm aggregates, the vertically drilled cores are approximately 8% stronger than comparable horizontally drilled specimens relative to cubes. This is as anticipated for larger specimens, but the measured strengths are approximately 10% stronger than expected from the Concrete Society recommendations (25), resulting in a lower correction factor to obtain an equivalent cube strength.

Table 5.1 Cube/corrected core conversion factors for 44 mm cores with $\lambda = 2.0$ (ref. 127)

Core orientation		Maximum aggregate size		
		10 mm	20 mm	Combined
Vertical	Conversion factor to 100 mm cube	1.05	1.25	1.15
	95% confidence limit on predicted cube strength (4 cores)	±17%	±23%	±23%
Horizontal	Conversion factor to 100 mm cube	1.14	1.22	1.17
	95% confidence limit on predicted cube strength (4 cores)	±15%	±17%	±17%

With 20 mm maximum aggregate, however, the cores were considerably weaker relative to cubes, confirming the influence of the aggregate size/core diameter ratio discussed above. In this case the orientation effect could not be detected. It is suggested that 10-mm and 20-mm aggregate concrete should be treated separately when converting 44-mm cores to equivalent cube strength. If this is done, the 95% confidence limits on the average of the results of groups of four cores of this size under laboratory conditions are unlikely to be better than the values given in Table 5.1. These may be approximated by $\pm 36/\sqrt{n}\%$ when n is the number of cores in the group. Bowman's reported results (128) also show a 7% higher strength for 50 mm cores when compared with 150 mm cores, but the aggregate size is not indicated. The Concrete Society (25), however suggests that strength differences between 'large' and 'small' cores are negligible and recommend the use of the formulae in section 5.2.2 for cores of 50 mm diameter and greater.

5.3.2 Reliability, limitations and applications

The reliability of compressive tests on small diameter cores is known to be less than for 'normal' specimens, and the authors have suggested a factor of 3 applied to the 95% confidence limits of predicted actual cube strengths under laboratory conditions. This gives a value of $\pm 36/\sqrt{n}\%$ for n cores with an aggregate size/diameter ratio of less than 1:3. But if the ratio of aggregate size/diameter is greater than 1:3, this accuracy is likely to decrease and may be as low as $\pm 50/\sqrt{n}\%$. Site cutting difficulties may further reduce accuracy. All the procedures described in section 5.1 concerning location, drilling and testing must be followed, just as for larger cores, and the effects of excess voidage and moisture accounted for as described in section 5.2. Small cores containing reinforcement should not be tested. Particular care must be taken to ensure that the core is representative of the mass of the concrete, and this is particularly critical in slabs drilled from the top surface in view of the reduced drilling depth required for a small core. BS 1881: Part 120 (114) does not refer to particular concrete densities, but ASTM C42 (115) specifically includes lightweight concrete in the range 1600 to 1920 kg/m^3, as well as normal-weight concrete.

There is no doubt that for applications other than compressive strength testing, small cores offer many economical and practical advantages compared with larger specimens. These applications include visual examination (including materials and mix details, compaction, reinforcement location and sizing); density determination; other physical tests, including point load or gas pressure tests; and chemical testing. For compressive strength testing, the chief limitation is variability of results and consequent lack of accuracy of strength prediction, unless many more specimens are taken than would normally be necessary. At least three times the number of 'standard' cores is required to give comparable accuracy, but it can be argued that this would still require less drilling in many instances and permits a wider spread of sample location. It is clear that considerable differences of predicted cube strength will arise from use of the various calculation methods, and as for larger cores it is essential that agreement is reached between all parties, before testing, about the method to be used.

Bowman (128) has described a successful approach in which 50-mm cores were used for strength tests on cast in-place piles because of their cheapness and ease of cutting, but were backed up by 150-mm cores where results were on the borderline of the specification. Another common situation in which small cores may be necessary for strength testing is when the slenderness of the member does not permit a larger diameter from the point of view of continued serviceability or adequate length/diameter ratio (> 1.0). This will apply especially to prestressed concrete members. Although in such cases small diameters are inevitable it is essential that the engineer fully appreciates the limitations of accuracy that he may expect. It may be that some other non-destructive approach will yield comparable accuracies of strength prediction, according to the availability of calibrations, with less expense, time and damage.

6   Load testing and monitoring

Where member strength cannot be adequately determined from the results of in-situ materials tests, load testing may be necessary. The expense and disruption of this operation may be offset by the psychological benefits of a positive demonstration of structural capacity which may be more convincing to clients than detailed calculations. In most cases where load tests are used, the main purpose will be proof of structural adequacy, and so tests will be concentrated on suspect or critical locations. Static tests are most common but where variable loading dominates, dynamic testing may be necessary.

Load testing may be divided into two main categories:

(i) In-situ testing, generally non-destructive
(ii) Tests on members removed from a structure, which will generally be destructive.

The choice of method will depend on circumstances, but members will normally only be removed from a structure if in-situ testing is impracticable, or if a demonstration of ultimate strength rather than serviceability is required. Ultimate strength capacity may sometimes be used as a calibration for other forms of testing if large numbers of similar members are in question.

Monitoring of structural behaviour under service conditions is an important aspect of testing which has received increased attention in recent years due to the growing number of older structures which are causing concern as a result of deterioration. This is considered in section 6.2 and many of the measurement techniques used for in-situ load testing and monitoring may also be useful for ultimate load test monitoring. There is also a growing trend towards monitoring the performance of older structures such as bridges during demolition to increase understanding of structural behaviour and the effects of deterioration (130, 131). Strain measurement techniques are described in section 6.3, and more specialized methods such as dynamic response and acoustic emission are included in Chapter 8.

6.1   In-situ load testing

The principal aim will generally be to demonstrate satisfactory performance under an overload above the design working value. This is usually judged by measurement of deflections under this load, which may be sustained for a specified period. The need may arise from doubts about the quality of construction or design, or where some damage has occurred, and the approach is particularly valuable where public confidence is involved. In Switzerland, static load tests on bridges are an established component of acceptance criteria and useful information concerning testing and deflection measurement procedures has been given by Ladner (132, 133). In other circumstances the test may be intended to establish the behaviour of a structure whose analysis is impossible for a variety of reasons. In this case strain measurements will also be necessary to establish load paths in complex structures. Views on detailed test procedures and requirements vary widely, but some commonly adopted methods are described in the following sections, together with suitable loading and monitoring techniques. Further guidance on general principles and basic procedures is provided by the Institution of Structural Engineers (5), whilst simple guidelines for static load tests on building structures have been given by Moss and Currie (134). Jones and Oliver (135) have also discussed some practical aspects of load testing, and Garas et al. (136) describe a number of more complex investigations. The philosophy of instrumentation of structures has also been considered by Menzies et al. (137). Practical difficulties of access and restraint will influence the preparatory work required, but in all circumstances it is essential to provide adequate safety measures to cater for the possible collapse of the member under test. The test loads will normally be applied twice, with the first cycle used for 'bedding-in' purposes.

6.1.1   Testing procedures
In-situ load tests should not be performed before the concrete is 28 days old unless there is evidence that the characteristic strength has been reached. ACI 318 (116) requires a minimum age of 56 days. Preliminary work is always necessary and must ensure safety in the event of a collapse under test, and that the full calculated load is carried by the members actually under test.

The selection of specific members or portions of a structure to be tested will depend upon general features of convenience, as well as the relative importance of strength and expected load effects at various locations. Attention must also be given to the parts of the structure supporting the test member. Selection of members may often be assisted by non-destructive methods coupled with visual inspection to locate the weakest zones or elements. Dynamic response approaches described in Chapter 8 may be useful in this respect.

6.1.1.1   Preliminary work.
Scaffolding must be provided to support at least twice the total load from any members liable to collapse together with the test load. This should be set to 'catch' falling members after a minimum drop, but at the same time should not interfere with expected deflections. Especial care must be taken to ensure that parts of the structure supporting such scaffolding are not overloaded in the event of a collapse under test, and that safeguards for unexpected failure modes (such as shear at supports) are provided.

The problem of ensuring that members under test are actually subjected to the assumed test load is often difficult, due to load-sharing effects. This may be a particular problem with floors or roofs supported by beams which span in one direction only. Even non-structural elements such as roofing boards may distribute loads between otherwise independent members, and in composite construction the effect becomes even greater. Whenever possible the member under test should be isolated from the surrounding structure. This may be achieved by saw cutting, although this is an expensive, tedious and messy operation. There will be many situations where this is not feasible from the point of view of reinstatement of the structure after test, or due to practicalities of load application. In such cases, test loads must be applied over a sufficiently large part of the structure to ensure that the critical members carry the required load. Load sharing characteristics are very difficult to assess in practice, but it is recommended (138) that in the case of beams with infill blocks and screed (Figure 6.1) the loaded width must be equal to at least the span to ensure that the central member is correctly loaded. It will be clear that this may lead to the need to provide very large test loads.

It will frequently be more convenient to concentrate loading above the member or group of members under examination, and to monitor relative deflections between this and all other adjacent members within a corresponding width. This will enable the proportion of load transferred away from the test member to be estimated, and the applied load can then be increased accordingly. It must be recognized that this increase may be between 2 and 4 times, and the shear capacity must be carefully checked.

Moss (139) has provided useful data for beam and block floors which includes thermal loading effects together with the influence of grouting and different screed types. Guidelines for assessing load-sharing effects, calculation of increased test loads to compensate and reductions of maximum allowable

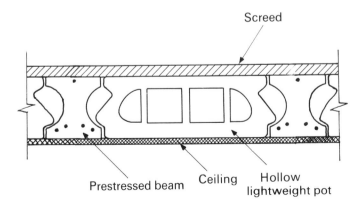

Figure 6.1 Beam and pot construction.

deflection criteria are all provided. Precautions must also be taken to ensure that members under test are not inadvertently supported by non-structural elements, such as partitions or services, although permanent finishes on the member need not be removed. Provision of a constant datum for deflection measurements is essential and must also be considered when carrying out preliminary work.

6.1.1.2   *Test loads.*   Views on test loads, which should always be added and removed incrementally, vary considerably. BS 8110 (140) requires that the total load carried should not be less than the sum of the characteristic design dead and imposed loads, but should normally be the greater of:

design dead load + 1.25 design imposed load

or 1.125 (design dead load + design imposed load)

(If any final dead load is not in position, compensating loads should be added.)

The test load is applied at least twice, with at least one hour between tests and with 5-minute settling time after the application of each increment before recording measurements. A third loading, sustained for 24 hours, may be useful in some cases. Performance is based initially on the acceptability of measured deflection and cracking in terms of the design requirements coupled with examination for unexpected defects. If significant deflections occur, the deflection recovery rates after removal of loading should also be examined. This deflection limit is not specified, but a value of $40l^2/h$ mm is sometimes used where l is the effective span in metres and h the overall depth in mm. Percentage recovery after the second test should not be less than that after the first load cycle, nor less than 75% for reinforced or partially prestressed concrete. Class 1 and 2 prestressed concrete members must satisfy a corresponding recovery limit of 85%.

ACI 318 (116) has similar provisions but with a test load sustained for 24 hours such that

total load = 0.85 (1.4 dead load + 1.7 imposed load).

Any shortfall of dead load should be made up 48 hours before the test starts, and the maximum acceptable deflection under test loads is given by:

$$\text{deflection limit} = \frac{(\text{effective span})^2}{20\,000 \times \text{member depth}}\text{(inches)}.$$

If this limit is exceeded, recovery must be checked.

The recovery limit for prestressed concrete is 80% but this may not be retested, whilst reinforced concrete failing to meet the 75% recovery criterion may be tested after 72 hours unloaded, but must achieve 80% recovery of deflections caused by the second test load.

The ACI requirements are 15–20% more stringent in terms of test load than BS 8110, and it is felt by many engineers that a total overload of only $12\frac{1}{2}\%$ is inadequate when certifying the long-term safety of a structure. The Institution of Structural Engineers (138) recognize this in recommending

$$\text{total load} = 1.25 \text{ (dead load + imposed load)}$$

when testing high alumina cement concrete structures. In such situations, where future deterioration is predicted, an even higher load may be justified. Lee (141) has proposed that a load of 1.5 (dead load + imposed load) should be adopted in all cases, but with greater emphasis on recording and analysing the members' response by means of load/deflection plots. Figure 6.2 shows a typical plot for an under-reinforced beam; experience would be required to recognize impending failure in order to stop load application.

6.1.2 *Load application techniques*
These are governed almost entirely by the practicalities of providing an adequate load as cheaply as possible at locations which are often difficult to access. The rate of application and distribution of the load must be controlled, and the magnitude must be easily assessed.

Water, bricks, bags of cement, sandbags and steel weights are amongst the materials which may be used and the choice will depend upon the nature and magnitude of load required as well as the availability of materials and ease of access. Care must be taken to avoid arching of the load as deflections

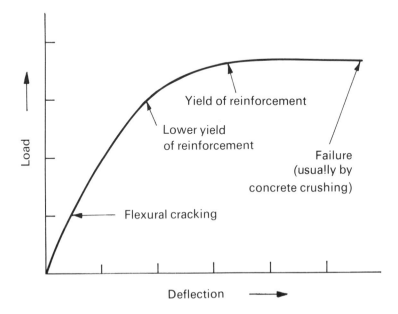

Figure 6.2 Typical load deflection curve for under-reinforced beam.

increase, and also unintended loads, such as rainwater, or those due to moisture changes of the loading material. In most cases a load which is uniform along the member length is required, but frequently this must be concentrated over a relatively narrow strip above the member under test. Steel weights, bricks or bags of known weight are best in this situation, and if the test member has been isolated it will probably be necessary to provide a platform clear of the adjacent structure (Figure 6.3). Figure 6.4 shows an alternative arrangement which may sometimes be more convenient for light roof purlins.

When loading is to be spread over a larger area, water may be the most appropriate method of providing the load. Slabs may be ponded by providing suitable containing walls and waterproofing, although care must be taken to allow for cambers or sags in calculating the loads. The effect may be reduced by baffling to create separate pools, but the likelihood of damage to finishes by leakage is high whenever ponding is used. An alternative to ponding is to provide containers such as plastic bins appropriately located, which can then be filled by hose to predetermined depths. Water is particularly useful in locations with limited space or difficult access, because storage and labour requirements will be reduced. Figures 6.5, 6.6 and 6.7 show some typical test loads applied to slabs and beams in 'building' structures. Test loads for bridges may often be conveniently provided by a suitable distribution of loaded wagons of known weight, such as water-filled truck mixers.

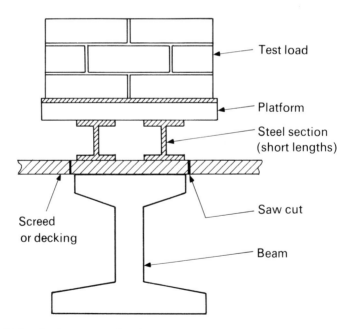

Figure 6.3 Test load concentrated on beam.

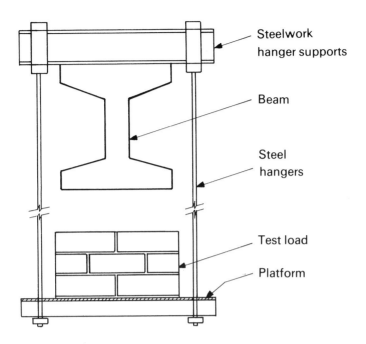

Figure 6.4 Test load arrangement for purlins.

For safety reasons, personnel working in a test load area must be restricted to those essential for load application and taking of measurements. Loads will always be applied in predetermined increments and in a way which will cause as little lack of symmetry or uniformity as possible. Similar precautions should be taken during unloading, and particular care is necessary to ensure that test load storage areas are not inadvertently overloaded. Deflection gauges must be carefully observed throughout the loading cycle, and if there are signs of deflections increasing with time under constant load, further loading should be stopped and the load reduced as quickly as possible. The potential speed of load removal is thus an important safety consideration, and 'bulk' loads which rely heavily on either manual or mechanical labour suffer the disadvantage of being relatively slow to handle. Water may be dispersed quickly if ponded, by the provision of 'knock-out' areas in the containing dyke, but the resultant damage to finishes may be considerable.

Non-gravity loading offers advantages of greater control, which can also be effected at a distance from the immediate test area, but is usually more expensive, and is restricted to use with load tests of a specialized or complex nature. Hydraulic systems may employ soil anchors, ballast or shoring to other parts of the structure to provide a reaction for jacking. The loading may be rapidly manipulated, and offers the advantage of simple cycling if required. In cases where a horizontal test load is required this approach may

Figure 6.5 Test load on roof slab using bricks (photograph by courtesy of Tysons Contractors Ltd.).

Figure 6.6 Test load on isolated beam using bricks (photograph by courtesy of Professor F. Sawko).

Figure 6.7 Test load concentrated over roof beam using steel weights (photograph by courtesy of Professor F. Sawko).

also be useful. Another apparently successful technique, described by Guedelhoefer (142), involves the application of a vacuum. Polythene-lined partition walls and seals must be constructed under the test area so that a vacuum can be drawn there by suction pump. This would be particularly suitable for slabs which cannot be loaded from above, or when a normal load is required on curved or sloping test surfaces. A maximum field-test pressure of $19.2 \, kN/m^2$ is claimed.

6.1.3 *Measurement and interpretation*
The measurement techniques associated with simple in-situ load tests are usually very straightforward, and are restricted to determination of deflections and possibly crack widths. Occasionally, more detailed results concerning strains and stress distributions will be required from a load test, or it may be necessary to monitor long-term behaviour of a structure under working conditions. The measurement techniques used here will be more complex, and are described separately below.

Basic in-situ load tests are based on deflection measurements, and these will normally be made by mechanical dial gauges which must be clamped to an independent rigid support. If scaffolding is used for this purpose, care must be taken to ensure that readings are not disturbed when the weight of the person taking the readings comes on to the scaffold. Dial gauges are often preferred to electronic or electric displacement transducers because a quick visual assessment of the progression of a load test is essential. However, use of a combined dial gauge/displacement transducer (Figure 6.8) will enable a visual on-site capability together with a complete data logging of the load test displacements for later retrieval, processing and presentation. Gauges will normally be located at mid-span and $\frac{1}{4}$ points (Figure 6.9) to check symmetry of behaviour. If the member is less than 150 mm in width, one gauge located on the axis at each point should be adequate, but pairs of gauges as in Figure 6.9 should be used for wider members. The selection of gauge size will be based on the expected travel, and although gauges can be reset during loading this is not recommended. The gauges must be set so that they can be easily read with a minimum of risk to personnel and so that the chance of disturbance during the test is small. Telescopes may often be convenient for this purpose. Readings should be taken at all incremental stages throughout the test cycles described in section 6.1.1.2 and temperatures

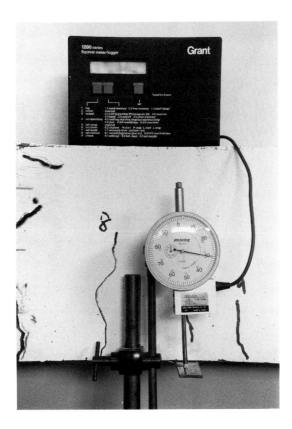

Figure 6.8 Dial gauge/displacement transducer with portable battery-powered logger.

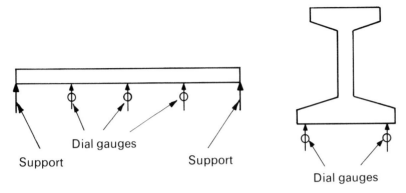

Figure 6.9 Gauge location.

noted at each stage. Measurement accuracy of $\pm\,0.1$ mm over a range of 6–50 mm is generally possible with dial gauges, and where sustained loads are involved it is prudent to have a back-up measurement system in case the gauges are accidentally disturbed. This will generally be less sensitive, and levelling relative to some suitable datum is probably the simplest method in most cases.

If the expected deflections are larger than can be accommodated by mechanical gauges, standard surveying techniques can be used in which scales are attached to the test element and monitored by a level. This has safety advantages, but the instrument must not be supported off the structure under test, which may cause difficulties. The accuracy expected will vary widely, but under normal conditions $\pm\,1.5$ mm may be possible (142). Shickert (14) has also described the use of laser holography for remote measurement of deflections.

Crack width measurements may occasionally be required, and these will normally be made with a hand-held illuminated optical microscope (Figure 6.10). This is powered by a battery and is held against the concrete surface over the crack. The surface is illuminated by a small internal light bulb and the magnified crack widths may be measured directly by comparison with an internal graduated scale which is visible through the eyepiece. A simple unmagnified comparitor scale can also be used (Figure 6.11) to assist in the estimation of crack widths.

Cracks present or developing in the course of a test should be traced on the surface with a pencil or coloured pen, with the tip marked at each load stage. Crack width readings should also be taken at each load stage at fixed locations on appropriate cracks. Visual identification of crack development may often be assisted by a surface coating of whitewash.

Interpretation of the results will often be a straightforward comparison of observed deflections or crack widths with limiting values which have been agreed previously between the parties concerned. Recommendations have

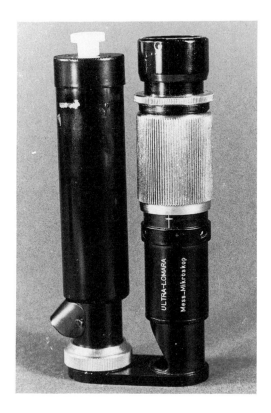

Figure 6.10 Crack microscope.

been outlined in section 6.1.1. The effects of load sharing will normally have been accounted for in the determination of the test load, so that the principal factor for which allowance must be made is temperature. This may cause significant changes of stress distributions within a member or structure between winter and summer, and differential of temperature across a member such as a roof beam may cause considerable deflection changes. Sometimes it may be possible to compensate for these by the establishment of a 'footprint' of movement for the structure for a range of temperatures (134).

As indicated previously, examination of the load/deflection plot can yield valuable information about the behaviour of the test member. A typical plot for a beam is given in Figure 6.12, in which the effects of creep during the period of sustained load and recovery can be seen. Since full recovery is not required instantaneously, some non-linearity of behaviour under sustained load is to be expected, but any marked non-linearity during the period of load application, other than that attributed to bedding-in and breakdown of non-structural finishes, must be taken as a sign that failure may be

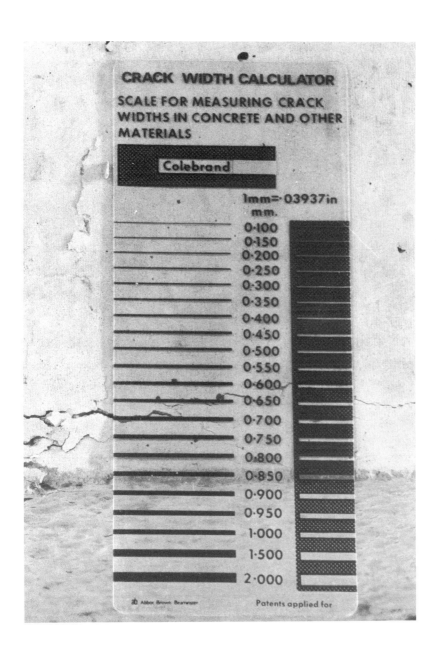

Figure 6.11 Crack width measurement scale.

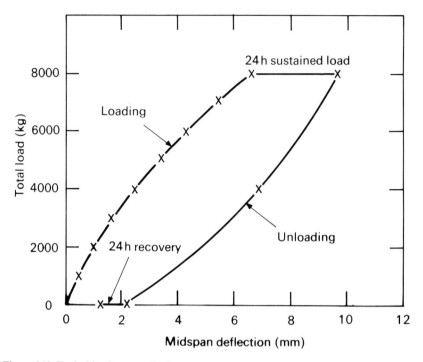

Figure 6.12 Typical load test result plot.

approaching. Comparison with Figure 6.2 may be useful, but if the member is over-reinforced, the warning of failure may be small, and experience is required to attempt to assess the reserve of strength.

6.1.4   *Reliability, limitations and applications*
The reliability of in-situ load tests depends largely upon satisfactory preparatory work to ensure freedom from unintended restraints, accurate load application to the test member, provision of an accurate datum for deflection measurements, and careful allowance for temperature effects. If these requirements are met, short-term tests should provide a reliable indication of the behaviour of the member or structure under the test load. This will be greatly enhanced by the examination of load/deflection plots as well as specific deflection values. Finishes or end restraints which are allowed for in design will frequently cause lower deflections than calculations would indicate, but providing these features are permanent this is not important.

   It is essential to recognize that a test of this type only proves behaviour under a specific load at one particular time. The behaviour under higher loads can only be speculative, and no indication of the margin of safety with respect to failure can be obtained. The selection of test load level must in most cases therefore be a compromise between continued serviceability of the test member and demonstration of adequate load-bearing capacity. If there is any possibility of future deterioration of the strength of materials, this must also be recognized in determining the test load together with any decisions based on the test results.

For short-term static tests instrumentation will generally be simple, but for long-term or dynamic tests, measurement devices must be carefully selected, particularly where strains are to be monitored. Gauges may be liable to physical damage or deterioration as well as temperature, humidity and electrical instability, and must also be selected and located with regard to the anticipated strain levels and orientation, as well as the gauge lengths and accuracies available.

Cost and inconvenience are serious disavantages of in-situ load testing, but these are usually outweighed by the benefits of a positive demonstration of ability to sustain required loads. Applications will therefore tend to be concentrated on critical or politically 'sensitive' structures, as well as those where lack of information precludes strength calculations. In-situ load tests are most likely to be needed in the following circumstances:

(i) Deterioration of structures, due to material degradation or physical damage.
(ii) Structures which are substandard due to quality of design or construction.
(iii) Non-standard design methods which may cause the designer, building authorities or other parties to require proof of the concept used.
(iv) Change in occupancy or structural modification which may increase loadings. Analysis is frequently impossible where drawings for existing structures are not available, but in other cases an inadequate margin may exist above original design values.
(v) Proof of performance following major repairs, which may be politically necesary in the case of public structures such as schools, halls and other gathering places.

In many cases chemical, non-destructive or partially-destructive methods may have been used as a preliminary to load testing to confirm the need for such tests, and to determine the relative materials properties of comparable members. Although estimates of the ultimate structural strength may be possible in the cases of deterioration of materials or suspect construction quality, a carefuly planned and executed in-situ load test will provide valuable information as to satisfactory behaviour under working loads.

Long-term monitoring of behaviour under service conditions will normally only be necessary if there is considerable doubt about future performance, or time-dependent changes are expected. This may occur if deterioration is expected, loadings are uncertain, or load testing is impracticable and there is a lack of confidence in strength estimates based on other methods.

6.2 Monitoring

This may range from relatively short-term measurements of behaviour during construction to long-term observations of behaviour during service.

6.2.1 Monitoring during construction

There has been a growing awareness of the need to monitor factors such as heat generation or strength development during construction of critical structures or parts of structures. Techniques for this purpose are outlined in other parts of this book. Interest has also grown in the implications of the use of newly developing high performance materials upon stresses and structural performance and some reports of monitoring are available (143).

6.2.2 Long-term monitoring

In cases of uncertainty about future performance of an element or structure it may be necessary to undertake regular long-term monitoring of behaviour. The scope of this ranges across the following:

    (i) modifications to existing structures
   (ii) structures affected by external works
  (iii) behaviour during demolition
  (iv) long-term movements
   (v) degradation of materials
  (vi) feedback to design
 (vii) fatigue assessment
(viii) novel construction system performance.

Moss and Matthews (144) have comprehensively reviewed these applications as well as practical issues concerning instrumentation and techniques to be used. A wide range of techniques are described ranging from simple visual and surveying methods, often supplemented by automatic data collection systems, to remote laser sensing systems to detect movements. Automation of data collection and storage is seen as a key aspect of many long-term monitoring situations. It is essential that measurements are taken at regular times each year to allow for seasonal effects, which may be considerable. Temperature, and preferably also humidity, should always be measured in conjunction with regular testing. Deflections and crack widths will normally be monitored, although strain measurements may also be valuable, particularly where the structure is subject to repetitive loading. In cases where there are large-scale movements of one part of a structure relative to another, permanent reference marks may be established and measured by rule, or if the area is normally not visible, a scale and pointer may be firmly fixed to the adjacent elements to indicate relative movement. This is a simple approach: more refined measurement systems will be required where movements are small.

Deflections may be monitored by levelling, using conventional surveying techniques, and this will detect major movements. Levels should always be taken at permanently-marked points on test members, preferably on a stud firmly cemented to the surface. Midspan readings should be related to readings taken as close as possible to the supports for each test member to determine deflection changes. Variations of this approach include taut string lines or piano wires stretched between the datum points at the supports, with a scale attached to the midspan point. A refinement involving a laser beam provides a means of detecting very small movements and may be appropriate for critical structures or members. This is suitable for continuous measurement, and may also be adapted to trigger an alarm system by replacing the scale with a suitable hole or slot through which the beam passes. A light-sensitive target would then react if the beam is cut off by excessive movement of the hole or slot (Figure 6.13). A similar approach using an infrared transmitter and receiver has also been described whilst another automatic system uses a taut stainless steel wire with proximity switches fixed to the beams to detect relative movement. Temperature-compensating springs maintain the tension, and a wire-break indicator is included in the control system which can provide an audible or light warning when a switch is triggered. A maximum wire length of 50 metres, with a control box handling up to 30 rooms, each containing 50 switches, has been claimed. Remote sensing photogrammatic techniques are also available (145).

Electrical variable-resistance potentiometers or displacement transducers may provide a useful method of accurate deflection measurement, particularly if automatic recording is an advantage. These consist of a spring-loaded centrally located plunger which is connected to a slide contact and moves up and down the core of a long wound resistor, usually 50–100 mm in length,

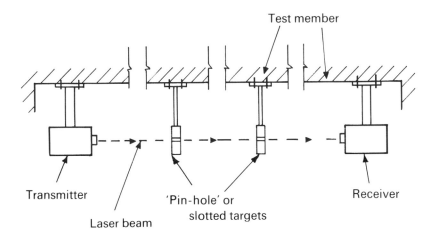

Figure 6.13 Laser beam for long-term monitoring.

and a voltage is applied to the system and recorded by simple digital voltmeter. It is claimed (142) that accuracies of 0.025 mm can be achieved in this way. The problem with long-term deflection measurements involving this type of equipment lies in providing suitable independent rigid supports and protection for the equipment, together with the long-term stability of calibration of the devices.

Crack widths may be measured using the optical equipment described in section 6.1.3, but this is inevitably subject to operators' judgements. A simple method of detecting the continued widening of established cracks is the location of a brittle tell-tale firmly cemented to the surface either side of the crack which will fracture if widening occurs. Thin glass slips, such as microscope slides, are commonly used but calibrated plastic tell-tales are also available. A mechanical method, using a Demec gauge as described in section 6.3, will provide direct numerical data on crack width changes and is simple to use if measuring studs are fixed permanently on either side of appropriate cracks. Equipment with a nominal 100 mm gauge length is recommended, and the same device must always be used for a particular test location.

If a precise measurement across cracks, or an automatic recording system, is required, the electrical displacement devices described above may be useful but their long-term performance should be carefully considered. Crack propagation may also be recorded automatically by electrical gauges consisting of a number of resistor strands connected in parallel and bonded to the concrete surface in a similar manner to strain gauges. As a crack propagates, individual strands within the high-endurance alloy foil grid will fracture and increase the overall electrical resistance across the gauge. This can be measured with a low voltage d.c. power supply and ohmmeter, and can be recorded automatically on a strip chart recorder. Additional low voltage instrumentation can be employed to trigger an alarm if this is necessary.

Strains may be measured by a variety of methods as described in section 6.3, but in most cases a simple mechanical approach using a Demec gauge will be the most successful for long-term monitoring. Provided that the studs are not damaged, readings may be repeated indefinitely over many years, and although the method suffers the disadvantage of a lack of remote readout, the problems of calibration and long-term 'drift' of electrical methods are avoided. If a remote reading system is preferred, vibrating wire gauges are generally regarded as reliable for long-term testing. The use of optical fibres to monitor strains is an important new development. This has become possible as a result of new technology and reported applications include monitoring crack widths in post-tensioned bridges as well as incorporation into prestressing tendons to monitor their in-service loads (144, 146). Two types of optical fibre are used:

(i) *Stranded optical fibre sensors.* Measurements can be made of light that

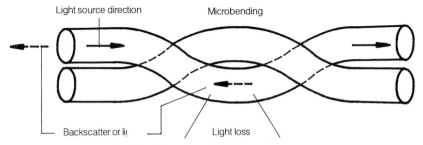

Figure 6.14 Stranded optical fibre (based on ref. 146).

is lost or attenuated whilst passing through regions containing microbends as illustrated in Figure 6.14. The intensity of light emerging as the sensor is strained or relaxed is compared with the light supplied, to enable changes in overall length to be made to an accuracy of ±0.02 mm for gauge lengths up to 30 m. This accuracy is independent of gauge length, but the precise position of localized straining is not given. Optical time domain reflectometry apparatus can be used to transmit pulses of light of about 1 ns duration into the sensor and measure the transit times of echoes from reflections at points of attenuation. This backscatter permits positions of attenuation to be located within ±0.75 m.

(ii) *Multi-reflection sensors.* These consist of single fibres containing up to 30 partial mirrors at intervals along their length. About 97% of the light passes through each partial mirror and the time for a pulse to be reflected back to the source is measured. The position of individual reflectors can be measured to an accuracy of ±0.15 mm using picosecond optical time domain reflectivity equipment.

Stranded optical fibres can easily be fitted to an existing structure to monitor long-term behaviour, and they are typically connected only at node points to enable strain distribution to be assessed. Multi-reflection sensors are particularly appropriate for incorporation into prestressing strands as a potentially 'intelligent' element in a 'smart' structure. Further practical details are given by Dill and Curtis (146), including information about control systems.

Vibrating wire gauges can be used for long-term monitoring in conjunction with a calibration load which is wheeled on a trolley across the floor or roof under test. This method is particularly suitable for monitoring slabs, and may be cheaper than making repeated readings which require scaffolding or ceiling removal for access.

Interpretation of long-term monitoring will normally consist of examination of load/deflection or load/crack width plots or the development of crack maps. Care must be taken to ensure that the effects of seasonal temperature and humidity changes are minimized or accounted for, and that loading conditions are similar whenever readings are taken for comparison. The importance of environmental records cannot be over-emphasized and factors such as central heating must not be overlooked. The frequency of testing will normally be determined according to the level of risk in conjunction with the observed trends of test readings with time.

6.3 Strain measurement techniques

Newly developed optical fibre techniques are described above, whilst established techniques for measurement of strain fall into four basic categories:

   (i) Mechanical
  (ii) Electrical resistance
 (iii) Acoustic (vibrating wire)
 (iv) Inductive displacement transducers.

The field of strain measurement is highly specialized, and often involves complex electrical and electronic ancillary equipment, and only the principal features of the various methods are outlined below, together with their limitations and most appropriate applications. Other methods which are of more limited value are also briefly described. The selection of the most appropriate method will be based on a combination of practical and economic factors, and whilst BS 1881: Part 206 (147) offers limited guidance, selection will largely be based on experience. Multidirectional 'rosettes' of gauges may be required in complex situations.

6.3.1 Methods available

The principal features of the available methods are summarized and compared in Table 6.1; the equipment and operation are described below.

   (i) *Mechanical methods.* The most commonly-used equipment consists of a spring lever system coupled to a sensitive dial gauge to magnify surface movements of the concrete. Alternatively a system of mirrors and beam of light reflected onto a fixed scale, or an electrical transducer may be used. A popular type, which is demountable and known as a Demec gauge (demountable mechanical), is shown in Figure 6.15.

Predrilled metal studs are fixed to the concrete surface by epoxy adhesive at a preset spacing along the line of strain measurement with the aid of a standard calibration bar. Pins protruding from the hand-held dial gauge holder, which comprises a sprung lever system, are located into the stud holes, enabling the distance between these to be measured to an accuracy of about 0.0025 mm. It is important that the Demec gauge is held normal to the surface of the concrete. 'Rocking' the gauge to obtain a minimum reading is recommended. Temperature corrections must be made with the aid of an

Table 6.1 Strain gauge summary

Type	Gauge length (mm)	Sensitivity (microstrain)	Special limitations	Advantages
Mechanical	12–2000	50–2	Contact pressure critical. Not recommended for dynamic tests	Cheap and robust Large gauge length
Electrical resistance (metal and alloy)	0.2–150	20–1	Low fatigue life. Surface preparation, temperature and humidity critical	Good for short gauge length
Semiconductor	19–100	1	Constant calibration required	Very accurate
Acoustic (vibrating wire)	12–200	10–1	Assembly and workmanship critical Avoid magnetic, physical and corrosive influences. Not recommended for dynamic tests	Good for long-term tests
Inductive displacement transducers	0.6–25	2–100	Expensive electrical ancillary equipment needed	Good for dynamic tests
Photoelastic			Does not measure strain accurately	Large strain range. Visual indication of strain configuration
Piezoelectric			Only suitable for laboratory use	Good for small rapid changes in strain

Figure 6.15 Photo of Demec gauge.

invar steel calibration bar, and readings may be repeated indefinitely provided the studs are not damaged or corroded. Cleanliness of the gauge and studs is important, and careful preparation of the surface is essential prior to fixing of the reference studs, which should be stainless steel for long-term testing.

The equipment is available in a range of gauge lengths, although 100, 200 and 250 mm are most commonly used. In general the greatest length should be used unless some localized feature is under examination, as in the case of laboratory model tests. The method is cheap, simple and the equipment relatively robust in comparison with other techniques, with the long gauge lengths particularly well suited to post-cracking testing of concrete and in-situ use. The principal disadvantage is the lack of a remote readout, and this can lead to very tedious operation where a large number of measurements are required. A further disadvantage is that different readings may be obtained by different operators.

A development of the method to give a remote readout is shown in Figure 6.16. This has a gauge length of 100 mm or 200 mm and uses Demec studs, with a standardized tension spring system to hold the equipment in position on the concrete surface. The flexible aluminium alloy strip is bent by relative movement of the pins, and this bending is measured by electrical resistance strain gauges. The equipment is particularly suited to laboratory use where a large gauge length is required. A similar approach has also been described by Din and Lovegrove (148) in which bending of a formed metal strip bolted to the concrete surface is monitored by electrical resistance gauges. This version is intended for long-term measurements where cyclic strains are involved.

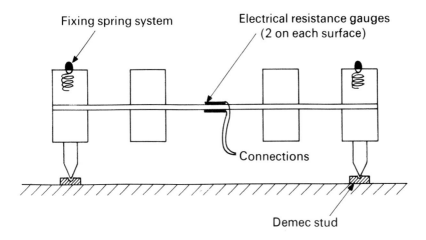

Figure 6.16 Portal strain transducer.

(ii) *Electrical methods.* The most common electrical resistance gauge is of the metal or alloy type in the form of a flat grid of wires, or etch-cut copper–nickel foil mounted between thin plastic sheets (Figure 6.17). This is stuck to the test surface, and strain is measured by means of changes in electrical resistance resulting from stretching and compression of the gauge. The resistance changes may be measured by a simple Wheatstone bridge which may be connected to multi-channel reading and recording devices. The relationship between strain and resistance will be approximately linear (for these gauges), and defined by the 'gauge factor'. Characteristics will vary according to the gauge construction, but foil gauges will generally be more sensitive, and have a higher heat dissipation which reduces the effects of self-heating.

The mounting and protection of gauges is critical, and the surface must be totally clean of dirt, grease and moisture as well as laitance and loose material. The adhesive must be carefully applied and air bubbles avoided, with particular care taken over curing in cold weather. These difficulties will tend to limit the use of such gauges under site conditions to indoor locations, where cotton wool placed over the gauge may provide a useful protective covering. Gauges may be cast into concrete if they are held in place and protected during concreting. They may also be fixed to the surface of reinforcing steel before or after casting but may cause local distortion of strains.

The relationship between strain and resistance will change with temperature and gauges may be self-compensating or incorporate a thermocouple. Alternatively a dummy gauge may be used to compensate for changes in the ambient temperature. Gauges must be sited away from draughts, although temperature will not affect readings over a small time scale of a few minutes.

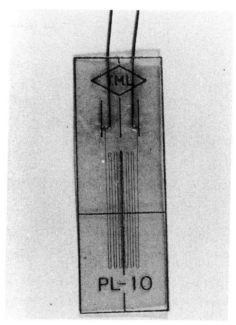

Figure 6.17 Electrical resistance strain gauge.

Humidity will also affect gauges, which must be water-proofed if they are subject to changes of humidity, or for long-term use. Background electrical noise and interference will also usually be present, and constant calibration is needed to prevent drift. If the gauge is subject to hysteresis effects these should be minimized by voltage cycling before use.

It is clear that the use of these gauges requires considerable care, skill and experience if reliable results are to be obtained. Their fatigue life is low, and this, together with long-term instability of gauge and adhesive, limits their suitability for long-term tests. The strain capacity will also generally be small unless special 'post-yield' gauges having a high strain limit are used. In the event that a surface-mounted strain gauge is located at the position of a subsequent tension crack, it is unlikely that a meaningful strain measurement will be obtained and the gauge may even be torn in half. Conversely, a gauge located immediately adjacent to a subsequent tension crack will not give representative results relating to the overall strain deformation of the concrete surface. Care therefore should be taken in using strain gauges on concrete surfaces that are expected to crack and the use of Demec or other gauges with a large gauge length is often preferable.

Electrical gauges with semiconductor elements are very sensitive, consisting of a grid fixed between two sheets of plastic, and are used in the same way as metal or alloy gauges. The same precautions of mounting, temperature control, and calibration apply, although they do not suffer from hysteresis effects. The gauges are very brittle however, requiring care in handling, and are unsuitable for casting into the concrete. The change in resistance is not directly proportional to strain and a precise calibration is therefore necessary, but the accuracy of measurement achieved is high.

(iii) *Acoustic (vibrating wire) methods.* These are based on the principle that the resonant frequency of a taut wire will vary with changes in tension. A tensioned wire is sealed into a protective tube and fixed to the concrete. An electromagnet close to the centre is used to pluck the wire, and it is then used as a pick-up to detect the frequency of vibration. This will normally be compared with the frequency of a dummy gauge by a recording device such as a cathode ray oscilloscope or comparative oscillator, to account for temperature effects. This type of gauge may be cast into the concrete (often encapsulated in precast mortar 'dog-bones'), and if adequately protected is considered suitable for long-term tests, although it is not appropriate for dynamic tests with a strain rate of greater than 1 microstrain/s (147) because of its slow response time. Particular care is necessary to avoid magnetic influences, and a stabilized electrical supply is recommended for the recording devices.

(iv) *Inductive displacement transducers.* Two series-connected coils form the active arms of an electrical bridge network fed by a high frequency a.c. supply. An armature moves between these coils, varying the inductance of each and unbalancing the bridge network. The phase and magnitude of the signal resulting from this lack of balance will be proportional to the displacement of the armature from its central position. The body of the gauge will be fixed to the concrete or a reference frame, so that the armature bears upon a metal plate fixed to the concrete. In this way lateral or diagonal as well as longitudinal strains may be measured; an adjusting mechanism allows zeroing of the equipment. These gauges are particularly sensitive over small lengths, but much expensive, complicated electrical equipment is needed to operate them and to interpret their output. This, together with the extensive precautions necessary, means that considerable experience is required.

(v) *Photoelastic methods.* These involve a mirror-backed sheet of photoelastic resin stuck to the concrete face. Polarized light is directed at this surface, and fringe patterns will show the strain configuration at the concrete surface under subsequent loading. The strain range of up to 1.5% is larger than any gauge can accommodate, but it is very difficult to obtain a precise value of strain from this method. The method may prove useful in examining strain distributions or concentrations at localized critical points of a member.

(vi) *Piezo-electric gauges.* The electrical energy generated by small movements of a transducer crystal coupled to the concrete surface is measured and related to strain. This is particularly suitable if small, rapid strain changes are to be recorded, since the changes generated are very short-lived, and these gauges are most likely to find applications in the laboratory rather than on site.

6.3.2 *Selection of methods*

The non-homogeneous nature of concrete generally excludes small gauge lengths unless very small aggregates are involved. For most practical in-situ testing, mechanical gauges will be most suitable, unless there is a particular need for remote reading, in which case vibrating wire gauges may be useful and more accurate over gauge lengths of about 100–150 mm. Both of these types are suitable for long-term use, given adequate protection. Optical fibres may be appropriate where measurements are required over considerable lengths.

Electrical resistance gauges may be useful if reinforcement strains are to be monitored, or for examining pre-cracking behaviour in the laboratory, and are usually associated with smaller gauge lengths, offering greater accuracy than mechanical methods. These gauges are, however, not re-usable; mechanical gauges have the advantage of not being damaged by crack formation across the gauge length. Semi-conductor gauges are very accurate, although delicate, and are likely to be restricted to specialized laboratory usage. For dynamic tests, electrical resistance or transducer gauges will be most suitable, although the operation of the latter will generally be more complicated and expensive. Photoelastic and piezo-electric methods have their own specialist applications outlined above. A great deal of information concerning the wide range of commercially available gauges including their maximum strain capacity is to be found in literature supplied by various equipment manufacturers, and it is recommended that this should be consulted carefully before selection of a particular gauge.

6.4 Ultimate load testing

Apart from its use as a quality control check on standard precast elements, ultimate load testing is an uncommon but important approach when in-situ overload tests are impossible or inadequate. The effort and disruption involved in the removal and replacement of members of a completed structure are considerable, but the results of a carefully monitored test, preferably carried out in a laboratory, provide conclusive evidence relating to the member examined. Comparison of the tested member with those remaining must be a matter of judgement, often assisted by non-destructive testing.

6.4.1 *Testing procedures and measurement techniques*

Ultimate load tests should preferably be carried out in a laboratory where carefully controlled hydraulic load application and recording systems are available. For small members it may be possible to use standard testing machines, and for larger members a suitable frame or rig can be assembled.

If the size of the test member prevents transportation to a laboratory it may be possible to assemble a test frame on site, adjacent to the structure from which the member has been removed. In this case load may be applied by manually operated jacks, with other techniques similar to those used in the laboratory. The control of load application and measurement will generally be less precise, however, and site tests should be avoided whenever possible.

6.4.1.1 Load arrangements. The most commonly used load arrangement for beams will consist of third point loading. This has the advantage of a substantial region of nearly uniform moment coupled with very small shears, enabling the bending capacity of the central portion to be assessed. If the shear capacity of the member is to be assessed, the load will normally be concentrated at a suitable shorter distance from a support.

Third point loading can be conveniently provided by the arrangement shown in Figure 6.18. The load is transmitted through a load cell or proving ring and spherical seating on to a spreader beam. This beam bears on rollers seated on steel plates bedded on the test member with mortar, high-strength plaster or some similar material. The test member is supported on roller bearings acting on similar spreader plates. Corless and Morice (149) have examined in detail the requirements of a mechanical test arrangement to avoid unwanted restraint and to ensure stability.

Details of the test frame will vary according to the facilities available and the size of loads involved, but it must be capable of carrying the expected test loads without significant distortion. Ease of access to the middle third for crack observations, deflection readings and possibly strain measurements is an important consideration, as is safety when failure occurs. Slings or stops may need to be provided to support the member after collapse.

In exceptional circumstances where two members are available it may be possible to test them 'back-to-back'. The ends can be clamped together with a spacer block by means of bolted steel frames, and the centres of the beams jacked apart by some suitable system incorporating a load cell. This method may be particularly suitable for tests conducted on site.

6.4.1.2 Measurements. Crack widths, crack development, and deflections will usually be monitored using the techniques described for in-situ testing. Dial gauges will be adequate for deflection readings unless automatic recording is required, in which case electrical displacement transducers may be useful. Strain measurements may not be required for straightforward strength tests, but are valuable for tests on prototype members or where detailed information about stress distributions is needed. The techniques available for strain measurement have been described in section 6.3, and the choice of method will involve many considerations, but the advantages of a large gauge length in the 'constant moment' zone are considerable.

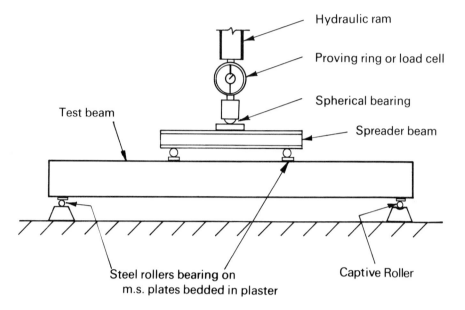

Figure 6.18 Laboratory beam load test arrangement.

6.4.1.3 Procedures. Before testing the member should be checked dimen-
sionally, and a detailed visual inspection made with all information carefully
recorded. If non-destructive tests are to be used for comparison with other
similar members, these should preferably be taken before final setting up in
the test frame.

After setting and reading all gauges, the load should be increased
incrementally up to the calculated working load, with loads, deflections and
strains, if appropriate, recorded at each stage. Cracking should be checked
visually, and a load/deflection plot prepared as the test proceeds. It will not
normally be necessary to sustain the working load for any specific period,
unless the test is being conducted as an overload test as described in section
6.1. The load should be removed incrementally, with readings again being
taken at each stage, and recovery checked. Loads will then normally be
increased again in similar increments up to failure, with deflection gauges
replaced by a suitably mounted scale as failure approaches. This is necessary
to avoid damage to gauges, and although accuracy is reduced, the deflections
at this stage will usually be large and easily measured from a distance.
Similarly, cracking and manual strain observations must be suspended as
failure approaches unless special safety precautions are taken. If it is essential
that precise deflection readings are taken up to collapse, electrical remote
reading gauges mounted above the test member may be necessary.

Modern load testing machines usually give the option of testing members

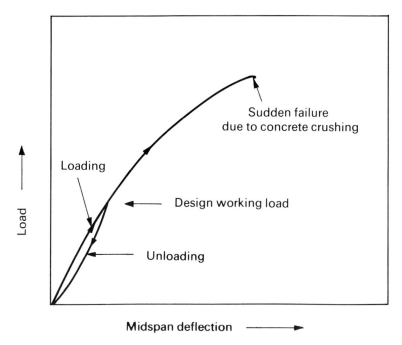

Figure 6.19 Load/deflection curve for typical over-reinforced prestressed beam.

under either 'load control' or 'displacement control'. The use of load control will result in a sudden catastrophic ultimate failure, as seen in Figure 6.19. The use of displacement control can enable the behaviour at the ultimate limit to be examined more carefully, so long as the overall stiffness of the load testing machine is significantly greater than that of the test member.

Crack development should be marked on the surface of the test member and the widths recorded as required. The mode and location of failure should also be carefully recorded — photographs, taken to show the failure zone and crack patterns, may prove valuable later. If information is required concerning the actual concrete strength within the test member, cores may be cut after the completion of load testing, taking care to avoid damaged zones.

6.4.2 Reliability, interpretation and applications

The procedures described above relate to basic load tests to determine the strength of a structural member. These are relatively straightforward, but the accuracy obtainable in a laboratory will be considerably higher than is possible on site. In specific situations, more complex tests may be required, involving dynamic loading, monitoring of steel strains, measurement of microcrack development or detailed strain distributions. Techniques suitable for such purposes are described elsewhere in this book, but are of a specialized nature. Testing of models or prototypes will also employ many of the techniques described but is regarded as being outside the scope of this book.

In most cases the results of load tests will be compared directly with strengths required by calculations, and serviceability limits required by specifications. The linear part of the load/deflection curve can be compared with the working loads carried by the member, and deflections and crack widths also compared with appropriate limits. The collapse load can be used to calculate the moment of resistance for comparison with required ultimate values. A typical load/deflection plot for an under-reinforced beam such as is shown in Figure 6.2 can also be used to determine the flexural stiffness. BS 8110 (140) requires that where tests are made on new precast units for acceptance purposes, the ultimate strength should be at least 5% greater than the design ultimate load and that the deflection at this load is less than 1/40 of the span.

Calculation of the collapse moment of resistance on the basis of tests on materials will not always provide a value which agrees closely with a measured value, as illustrated by Figure 6.20, which shows measured collapse loads for a series of pretensioned beams compared with predicted values calculated

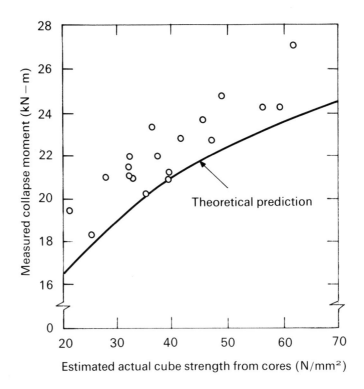

Figure 6.20 Measured vs. calculated collapse moments for prestressed concrete beams.

from small core strength estimates from the same beams. For over-reinforced beams, the dependence of collapse moment on concrete strength is likely to be greater (Figure 6.19). Particular care must be taken when testing over-reinforced beams unless displacement control is used as discussed above because of the sudden nature of failure, which often provides little warning from increasing deflections.

A test to destruction is obviously the most reliable method of assessing the strength of a concrete member, but the practical value of tests on members from existing structures depends upon the representativeness of the member tested. Visual and non-destructive methods may be used for comparisons, but uncertainty will always remain. The most likely application, apart from quality control checking of precast concrete, will be where a large number of similar units are the subject of doubt. The sacrifice of some of these may be justified in relation to the potential cost of remedial works.

7 Durability tests

Deterioration of structural concrete may be caused either by chemical and physical environmental effects upon the concrete itself, or by damage resulting from the corrosion of embedded steel. The tests described in this chapter are concerned primarily with assessment of material characteristics which are likely to influence the resistance to such deterioration, and to assist identification of the cause and extent if it should occur. These tests have been summarized in Table 1.3, although those involving chemical or petrographic analysis (including carbonation depth, and sulphate and chloride content) are considered in detail in Chapter 9. Other relevant tests relating to structural integrity and performance are described in Chapter 8, and test selection is discussed in section 1.4.3 of Chapter 1.

The principal causes of degradation of the concrete are sulphate attack, alkali-aggregate reaction, freeze–thaw, abrasion and fire. Reinforcement corrosion is an electro-chemical process requiring the presence of moisture and oxygen and can only occur when the passifying influence of the alkaline pore fluids in the matrix surrounding the steel has been destroyed, most commonly by carbonation or chlorides (150). The presence of moisture and its ability to enter and move through the concrete are thus critical features since both sulphates and chlorides require moisture for mobility and alkali-aggregate reactions cannot occur in dry concrete. Carbonation rates depend on gas permeability and are also influenced by moisture levels. Tests which assess water and gas absorption or permeability, and moisture content, are thus of great importance with respect to durability. Planning and interpretation of a typical corrosion-related investigation are outlined in Appendix A7.

7.1 Corrosion of reinforcement and prestressing steel

Corrosion of embedded steel is probably the major cause of deterioration of concrete structures at the present time. This may lead to structural weakening due to loss of steel cross-section, surface staining and cracking or spalling. In some instances internal delamination may occur. The corrosion process has been described in detail by many authors but is summarized in simple form by Figure 7.1.

This may involve either localized 'micro-corrosion' cells in which pitting may severely reduce a bar cross-section with little external evidence, or generalized 'macro-corrosion' cells which are likely to be more disruptive and easier to detect due to expansion of the rusting steel (151).

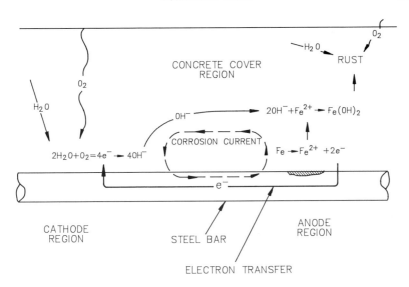

Figure 7.1 Basic corrosion process of steel reinforcement in concrete.

The development of anodic and cathodic regions on the surface of a steel reinforcing bar results in a transfer of ions within the concrete cover and of electrons along the bar and hence a flow of corrosion current. The rate at which corrosion occurs will be controlled either by the rate of the anodic or cathodic reactions or by the ease with which ions can be transferred between them. Thus an impermeable concrete, normally associated with a high electrical resistivity, will restrict ionic flow and hence result in low rates of corrosion. A thick and impermeable cover region will also restrict the availability of oxygen to the cathode region and further reduce the rate of corrosion.

The presence of corrosion activity can often be detected by measuring the electrochemical potentials on the surface of the concrete with respect to a reference half-cell. The test can be used in conjunction with electrical resistivity measurements of the concrete cover zone to give an indication of the probable rate of corrosion activity. Alternatively, the rate of corrosion may also be measured directly by a number of perturbative electrochemical techniques. The most popular of these is the linear polarization resistance measurement. A simple measurement of the thickness of the concrete cover will also give some guidance to the expected durability of a reinforced concrete structure.

7.1.1 *Electromagnetic cover measurement*

Electromagnetic methods are commonly used to determine the location and cover to reinforcement embedded in concrete. Battery-operated devices commercially available for this purpose are commonly known as cover-meters. A wide range of these is commercially available and their use is covered by BS 1881: Part 204 (152).

7.1.1.1 Theory, equipment and calibration. The basic principle is that the presence of steel affects the field of an electromagnet. This may take the form of an iron-cored inductor of the type shown in Figure 7.2. An alternating current is passed through one of the coils, while the current induced in the other is amplified and measured. The search head may in fact consist of a single or multiple coil system, with the physical principle involving either eddy current or magnetic induction effects. Eddy current instruments involve measurement of impedance changes and will be affected by any conducting metal, and magnetic induction instruments involve induced voltage measurements and are less sensitive to non-magnetic materials.

The influence of steel on the induced current is non-linear in relation to distance and is also affected by the diameter of the bar, which makes calibration difficult. Commonly used simple types of covermeter (Figure 7.3) overcome this problem by using two ranges for cover, typically 0–40 mm and 40–100 mm. The calibrated scale is marked in bands associated with varying cover, and this accommodates the influence of reinforcement diameter. Small bars will give a reading at the 'high' end but large bars will give a reading at the 'low' end of a particular band, because the effect of diameter is relatively small over the range of bar sizes from 10–32 mm. If bars of less than 10 mm or greater than 32 mm are to be measured, it may be necessary to develop a special calibration, and the linear scale which is usually provided

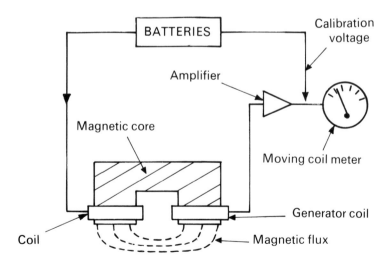

Figure 7.2 Typical simple covermeter circuitry.

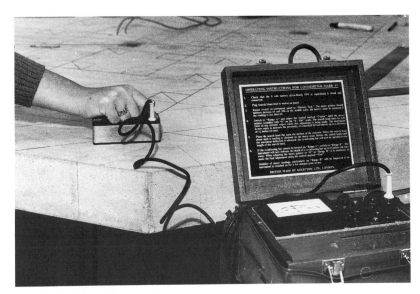

Figure 7.3 Typical simple covermeter.

can be used. More refined versions of this type of equipment, involving more sophisticated electronic circuitry and digital output, are also available, and are capable of allowing for bar diameter, and also of detecting bars at a greater depth (in some cases up to 300 mm). These are considerably more expensive than the basic equipment described above. One such version is illustrated in Figure 7.4. A microprocessor model which incorporates allowance for steel type and provides an audible 'low cover' warning facility is also available (Figure 7.5).

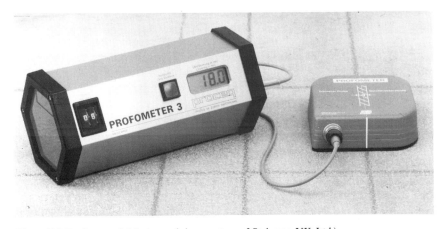

Figure 7.4 Profometer 3 (photograph by courtesy of Steinweg UK Ltd.).

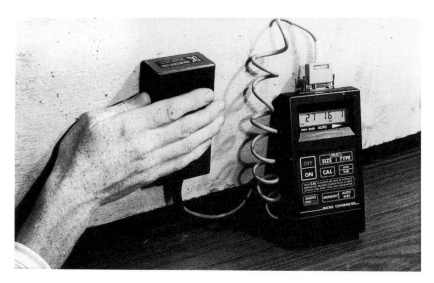

Figure 7.5 Microcovermeter.

Recent developments in covermeter equipment have resulted in several models that offer to evaluate both the cover to the bar and the bar diameter itself, where this is not known. This is accomplished either by the use of a spacer block (153) or by the use of a specialized search head (Figure 7.4). The ability to scan a covermeter over the concrete surface and continuously record the output into a data logger for subsequent graphical presentation has also recently become available.

Basic calibration of the equipment is important and BS 1881: Part 204 (152) suggests several alternative methods. These include the use of a test prism of ordinary Portland cement concrete. A straight clean reinforcing bar of the appropriate type is embedded to project from the prism and to provide a range of covers which can be accurately measured by steel rule for comparison with meter readings. Other methods involve precise measurements to a suitably located bar in air. In all methods it is essential that extraneous effects upon the magnetic field are avoided. Under these conditions the equipment should be accurate to $\pm 5\%$ or 2 mm, whichever is the greater.

Site calibration checks should also be performed with the bar and concrete type involved in the investigation. This may involve the drilling of test holes at a range of cover values to verify meter readings, and if necessary to reset the equipment or develop a separate calibration relationship.

Further development is expected to provide a new type of covermeter based upon the principle of magnetic flux leakage (154). A d.c. magnetic field normal to the axis of a reinforcing bar is set up via a surface yoke, which partially magnetizes the bar. A sensor moved from one pole of the yoke to the other detects the induced magnetic leakage field, which can be used to determine both the depth and diameter of the bar. Magnetic flux leakage also has the potential to enable detection of a reduction in the section of the reinforcing bar, such as might be caused by severe pitting corrosion. Efforts are being made to use neural network artificial intelligence to simplify interpretation of the results.

7.1.1.2 Procedure. Most covermeters consist of a unit containing the power source, amplifier and meter, and a separate search unit containing the electromagnet whch is coupled to the main unit by a cable. In use, the reading will be zeroed and the hand-held search unit moved over the surface of the concrete under test. The presence of reinforcement within the working range of the equipment will be indicated by movement of the indicator needle or digital value. The search unit is then moved and rotated to obtain a maximum reading and this position will correspond to the location of a bar (minimum cover). With some instruments this is assisted by a variable-pitch audible output. The needle or output will then indicate the cover on the appropriate scale, whilst the direction of the bar will be parallel to the alignment of the search unit. The use of spacers may also be necessary to improve the accuracy of cover measurements which are less than 20 mm.

7.1.1.3 Reliability, limitations and applications. Although the equipment can be accurately calibrated for specific reinforcement bars (section 7.1.1.1) in most practical situations the accuracy that can be achieved will be considerably reduced. The factors most likely to cause this are those which affect the magnetic field within the range of the meter, and include:

(i) Presence of more than one reinforcing bar. Laps, transverse steel as a second layer or closely-spaced bars (less than three times the cover) may cause misleading results. With some equipment a small, non-directional 'spot probe' can be used to improve discrimination between closely spaced bars and to locate lateral bars.
(ii) Metal tie wires. Where these are present or suspected, readings should be taken at intervals along the line of the reinforcement and averaged.
(iii) Variations in the iron content of the cement, and the use of aggregates with magnetic properties, may cause reduced covers to be indicated.
(iv) A surface coating of iron oxide on the concrete, resulting from the use of steel formwork, has been claimed to cause a significant underestimate of the reinforcement cover and should be guarded against.

BS 1881: Part 204 (152) suggests that an average site accuracy at covers less than 100 mm of about $\pm 15\%$ may be expected, with a maximum of ± 5 mm, and it is important to remember that the calibrated scales are generally based on medium-sized plain round mild steel bars in Portland cement concrete. If the equipment is to be used in any of the following circumstances a specific recalibration should be made:

   (i) Reinforcement less than 10 mm diameter, high tensile steel or deformed bars. In these cases the indicated cover is likely to be higher than the true value. This will also apply if the bars are curved and hence not parallel to the core of the electromagnet.

  (ii) Special cements, including high alumina cement, or added pigments. In these cases the indicated cover is likely to be lower than the true value.

 (iii) Reinforcement in excess of 32 mm diameter may require a recalibration for some models of covermeter.

Estimates of bar diameter may only be possible to within two bar sizes. The operating temperature range of covermeters is also generally relatively small, and battery-powered models will not usually function satisfactorily at temperatures below freezing, which may seriously limit their field use in winter. Stability of reading may be a problem with some types of instrument and frequent zero checking is essential.

The most reliable application of this method to in-situ reinforcement location and cover measurement will be for lightly reinforced members. As the complexity and quantity of reinforcement increases, the value of the test decreases considerably (155), and special care should also be taken in areas where the aggregates may have magnetic properties. Malhotra (50) has described an application to precast concrete quality checking, in which the linear scale is calibrated to enable an acceptable range of values to be established for routine component monitoring. Snell, Wallace and Rutledge (156) have also considered detailed sampling plans for in-situ investigations and developed a statistical methodology for such situations. Alldred (157) has compared a number of different covermeters on congested steel reinforcement and gives some correction factors that may be used to accommodate errors of measurement.

7.1.2   Half-cell potential measurement

This method has been developed and widely used with success in recent years where reinforcement corrosion is suspected, and normally involves measuring the potential of embedded reinforcing steel relative to a reference half-cell placed on the concrete surface as shown in Figure 7.6. ASTM C876 (158) covers this method.

7.1.2.1   Theory, equipment and procedures.   The reference half-cell is usually a copper/copper sulphate or silver/silver chloride cell but other combinations have been used (159). Different types of cell will produce different values of surface potential and corrections of results to an appropriate standardized cell may be necessary during interpretation. The concrete functions as an

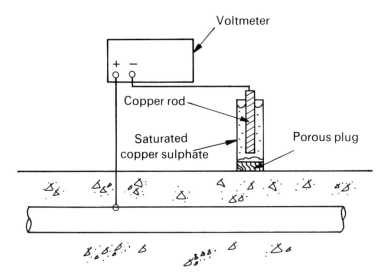

Figure 7.6 Reinforcement potential measurement.

electrolyte and anodic, corroding regions of steel reinforcement in the immediate vicinity of the test point may be related empirically to the potential difference measured by the high-impedance voltmeter.

It is usually necessary to break away the concrete cover to enable electrical contact to be made with the steel reinforcement. This connection is critical and a self-tapping screw is recommended, but adequate electrical continuity is usually present within a mesh or cage of reinforcement to avoid the need for repeated connections (160). Some surface preparation, including wetting, is also likely to be necessary to ensure good electrical contact. Two-cell methods, avoiding the need for electrical connections to reinforcement can be used for comparative testing (159), but are regarded as less reliable.

The basic equipment is very simple and permits a non-destructive survey of the surface of a concrete member to produce iso-potential contour maps as illustrated by Figure 7.7.

A range of commercially available equipment is available including digital single-reading devices (161) (Figure 7.8), as well as 'multi-cell' and 'wheel' devices (Figure 7.9) with automatic data logging and printout facilities designed to permit large areas to be tested quickly and economically (162).

7.1.2.2 Reliability, limitations and applications. Early studies on half-cell potentials (163) were primarily concerned with elevated bridge decks in the USA, where unwaterproofed concrete was treated every winter with large quantities of deicing salts. In conditions where there is a plentiful availability of oxygen and where chloride contamination is ingressing from the surface,

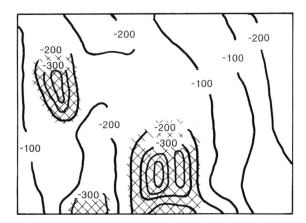

Figure 7.7 Typical half-cell potential contours (based on ref. 159). ⊠; zones requiring further investigation.

Figure 7.8 Single half-cell instrumentation.

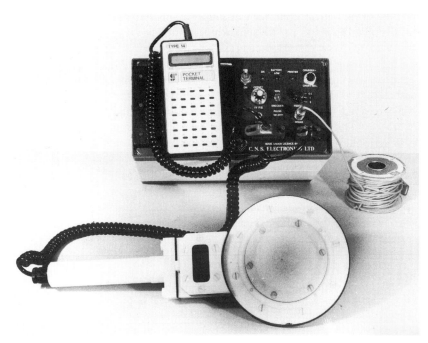

Figure 7.9 'Wheel' half-cell equipment.

interpretive guidelines can be given (Table 7.1) to assess the risk of corrosion occurrence. Care should be taken in applying those guidelines to different environmental conditions. Studies on European bridge decks (164) where waterproofing membranes are used or where deicing salts are applied less frequently have resulted in a different set of intepretive guidelines. Although a wet contact is needed between the half-cell and the concrete, a complete

Table 7.1 General guides to interpretation of electrical test results (based on refs 158 and 162)

Half-cell potential (mV) relative to copper/copper sulphate reference electrode	Percentage chance of active corrosion
< −350	90%
−200 to −350	50%
> −200	10%

Resistivity (ohm-cm)	Likelihood of significant corrosion (non-saturated concrete when steel activated)
< 5000	Very high
5000–10 000	High
10 000–20 000	Low/moderate
> 20 000	Low

wetting of the concrete surface can cause significantly more negative potential results by up to 200 mV (160, 165), whilst completely water-saturated concrete can lead to oxygen starvation, resulting in potential values more negative than −700 mV (166).

Studies on carbonated concrete have shown (167) that corrosion is typically associated with half-cell potential readings in the range −200 mV to −500 mV but with considerably more shallow potential gradients than seen with external chloride contamination.

Where chlorides are present in the concrete, due to the use of calcium chloride accelerator in the original mix, corrosion is typically associated with half-cell potential readings in the range +100 mV to −400 mV (167), with often a very narrow range of potentials separating rapidly corroding and, apparently, uncorroding areas. Half-cell potential gradients are frequently shallow, due to the close proximity of adjacent corrosion 'micro-cells' on the surface of the steel reinforcement.

In the light of these conflicting interpretive criteria, it is now more common not to use absolute potential values as a means of assessing the probability of corrosion occurrence. The plotting of iso-potential contour maps such as in Figure 7.7 is preferred. Local corrosion risk is identified by 'islands' of more negative, anodic regions and by steep potential gradients, seen by closely spaced iso-potential lines. Regions which are more negative than 200 mV from the 'background potential' are often indicative of corrosion activity (160).

In carrying out a potential survey, an initial grid of 0.5 m to 1 m is commonly used to sample the surface potentials. In regions of particular interest or where micro-cell corrosion activity is suspected, a grid as fine as 0.1 m may be used. It is essential to recognize that the half-cell method cannot indicate the actual corrosion rate or even whether corrosion has already commenced. The test only indicates zones requiring further investigation, and an assessment of the likelihood of corrosion occurring may be improved by resistivity measurements in these regions.

This method is widely used when assessing maintenance and repair requirements. It is particularly valuable in comparatively locating regions in which corrosion may cause future difficulties, and those in which it has already occurred but with no visible evidence at the surface. It can also often be of use to confirm that passivity has been restored following remediation to a corrosion-damaged reinforced concrete structure.

7.1.3 *Resistivity measurements*

The ability of corrosion currents to flow through the concrete can be assessed in terms of the electrolytic resistivity of the material. Procedures for in-situ measurement are available for use in conjunction with half-cell potential measurements, but at the present time the method is less widely used.

7.1.3.1 Theory, equipment and procedures. Electrical resistivity tests have been used for soil testing for many years using a Wenner four-probe technique, and this has recently been developed for application to in-situ concrete. Four electrodes are placed in a straight line on, or just below, the concrete surface at equal spacings as shown in Figure 7.10. A low frequency alternating electrical current is passed between the two outer electrodes whilst the voltage drop between the inner electrodes is measured. The apparent resistivity is calculated as

$$\rho = \frac{2\pi s V}{I}$$

where s is the electrode spacing, V is the voltage drop and I is the current. A spacing of 50 mm is commonly adopted and resistivity is usually expressed in ohm-cm or in k.ohm-cm.

Considerable efforts have been made to develop portable equipment which permits satisfactory electrical contact without the need to drill holes into the surface, and a device incorporating spring-loaded contacts with an automatic couplant dispenser has been introduced by Millard (168) (Figure 7.11). An alternative approach (169) has been to use a square wave a.c. current to accommodate the effects of a poor surface contact. Both techniques have

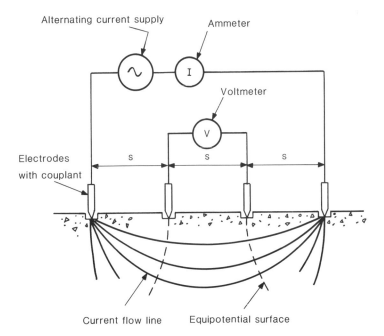

Figure 7.10 Four-probe resistivity test.

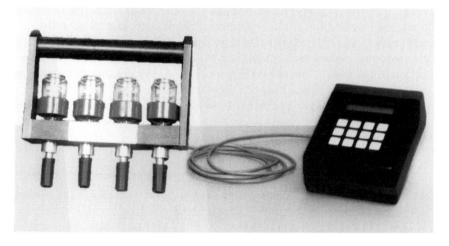

Figure 7.11 Four-probe resistivity equipment.

been shown (170) to give very similar results in most practical situations. This permits a rapid non-destructive assessment of concrete surface zones. Hand-held 'two-probe' equipment is also available although this requires holes to be drilled and experience with this approach is limited.

7.1.3.2 Reliability, limitations and applications. Classification of the likelihood of corrosion actually occurring can be obtained on the basis of the values in Table 7.1 when half-cell potential measurements show that corrosion is possible. The resistivity of concrete is known to be influenced by many factors including moisture and salt content and temperature as well as mix proportions and water/cement ratio. McCarter *et al.* (171) have shown from laboratory studies that resistivity decreases as the water/cement ratio increases and since the resistivity of aggregate may be regarded as infinite relative to that of the paste, the value for concrete is dependent upon the paste characteristics and proportion. It is also claimed that the resistivity may be used as a measure of the degree of hydration of the cement in the mix. Further fundamental studies of concrete resistivity have been reported by Wilkins (172).

For in-situ resistivity measurements there are a number of practical considerations that must be accommodated before interpreting the results.

(i)   The presence of steel reinforcement close to the measurement location will cause an underestimate in the assessment of the concrete resistivity (173).

(ii) The presence of surface layers due to carbonation or surface wetting can cause a significant underestimate or overestimate in the underlying concrete resistivity (174).

(iii) Taking resistivity measurements on a very small member section or close to a section edge may result in an overestimate of the actual resistivity (175).

(iv) Resistivity measurements will fluctuate with changes in ambient temperature and with rainfall. In external conditions in the UK, where concrete is normally moist, temperature is found to be the more dominant parameter (176).

Some efforts have been made (177, 178) to relate half-cell potential and resistivity measurements to rates of corrosion through computer modelling, but this approach has not yet seen significant practical usage.

Although the principal application is for assessment of maintenance and repair requirements in conjunction with half-cell potential measurements, the technique may be applied on a larger scale with greater spacings to estimate highway pavement thicknesses. Moore (179) has described US Federal Highway Administration work based on the differing resistivity characteristics of pavement concrete and subgrade. Electrode spacings are varied and a change of slope of the resistivity/spacing plot will occur as a proportion of the current flows through the base material.

7.1.4 *Perturbative measurement of corrosion rate*

A number of perturbative electrochemical techniques have been developed to determine directly the rate of corrosion by measuring the response of the corrosion interface to a small perturbation. The principal methods are:

(i) Linear polarization resistance measurement.
(ii) Galvanostatic pulse transient response measurement.
(iii) AC impedance analysis.
(iv) AC harmonic analysis.

These methods have been very little tried in the field to date in the UK, although linear polarization has seen some limited usage in the rest of Europe (180, 181) and in the USA (182) and the galvanostatic pulse method has been applied to some European bridges (183, 184).

Linear polarization resistance measurement is carried out by applying a small electrochemical perturbation to the steel reinforcement via an auxiliary electrode placed on the surface of the concrete (Figure 7.12). The perturbance is often a small d.c. potential change, ΔE, to the half-cell potential of the steel in the range $\pm 20\,\text{mV}$. From a meaurement of the resulting current, ΔI, after a suitable equilibration time, typically 30 seconds to 2 minutes, the polarization resistance, R_p, is obtained, where

$$R_p = \Delta E/\Delta I$$

R_p is inversely related to the corrosion current, I_{corr}, flowing between anodic and cathodic regions on the surface of the steel bar. Hence

$$I_{corr} = B/R_p,$$

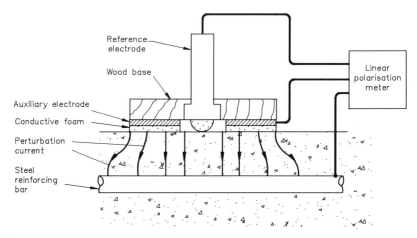

Figure 7.12 Linear polarization resistance measurement.

where for steel in concete B normally lies between 25 mV (active) and 50 mV (passive). The corrosion current density, i_{corr}, is found from

$$i_{corr} = I_{corr}/A$$

where A is the surface area of the steel bar that is perturbed by the test. Typical values of i_{corr} and the resulting rate of corrosion penetration are shown in Table 7.2 (185). A number of difficulties with this technique must be considered:

(i) The time for equilibration to occur may result in excessive measurement times or inaccurate corrosion rate evaluations if premature measurements are taken.

(ii) When the concrete resistance, R, between the auxiliary electrode on the surface of the concrete and the steel reinforcing bar is high, this can result in significant errors in measuring R_p unless R is either electronically compensated or explicitly measured and deducted from R_p.

(iii) The use of a correct value for B requires a foreknowledge of the corrosion state of the steel. Adopting an inappropriate value could result in an error of up to a factor of two.

Table 7.2 Typical corrosion rates for steel in concrete

Rate of corrosion	Corrosion current density, i_{corr} ($\mu A/cm^2$)	Corrosion penetration, p ($\mu m/year$)
High	$10 \rightarrow 100$	$100 \rightarrow 1000$
Medium	$1 \rightarrow 10$	$10 \rightarrow 100$
Low	$0.1 \rightarrow 1$	$1 \rightarrow 10$
Passive	< 0.1	< 1

(iv) It is tacitly assumed that corrosion occurs uniformly over the measurement area, A. Where localized pitting occurs then this assumption could result in a significant underestimate of the localized rate of corrosion.

(v) Evaluating the area of measurement A is not simple. Some studies (186) recommend using a large (250 mm diameter) auxiliary electrode and assume that the surface area of measurement is the 'shadow area' of steel reinforcement lying directly beneath the auxiliary electrode. An alternative approach (187) is to accept that the perturbation current will try to spread laterally to steel lying outside the shadow area and to confine this lateral spread by use of a guard ring positioned around the auxiliary electrode. Both techniques show considerable promise in the field, but the accuracy of each approach has yet to be independently verified.

(vi) Corrosion rates measured on one occasion may not be typical of the mean annual rate of corrosion. Fluctuations in ambient temperature, moisture levels in the concrete, oxygen availability, etc. may all cause the instantaneous corrosion rate to vary significantly. Only by taking a series of measurements under different environmental conditions can an overall evaluation of the annual rate of corrosion be made.

The galvanostatic pulse transient response method uses a surface electrode arrangement similar to the linear polarization method (Figure 7.12). A small current perturbation, ΔI, is applied to the steel reinforcing bar and the resulting shift in the half-cell potential, ΔE, is measured. The transient behaviour of ΔE can be used to evaluate the concrete resistance, R, and to evaluate the rate of corrosion, I_{corr} (188). Equipment specifically designed for field application of the galvanostatic pulse transient response method has not yet been commercially developed but the technique holds considerable promise.

Both the a.c. impedance and the a.c. harmonic analysis techniques are laboratory perturbative electrochemical methods that can evaluate the instantaneous rate of corrosion. However, neither of these methods has seen the field developments that have been applied to the linear polarization and galvanostatic pulse transient response techniques. Their complexity makes it less likely that they will ever be used in the field for routine corrosion measurements.

7.1.5 Other electrical techniques

Some studies (189) have related small fluctuations (0.1 to 10 mV) in the half-cell potential of the reinforcing steel to the spontaneous formation and repassivation of corrosion anodes. These fluctuations are called 'potential noise' and can be related to the rate of corrosion. This technique looks promising but has yet to see extensive use in the field.

A method of embedding a series of mild steel samples at increasing depth has been used (190) to monitor the ingress of carbonation or chlorides and to evaluate the influence upon the rate of corrosion. Effective use of this technique requires either planning before construction of the structure or subsequent embedding into a void cut into the hardened concrete. The latter technique may place the sensors in an environment unrepresentative of the rest of the structure and care should be used in implementing this approach.

7.2 Moisture measurement

Measurement of the internal moisture content in concrete is surprisingly difficult, but nevertheless of considerable importance when assessing durability performance. Alkali-aggregate reactions are moisture dependent whilst the effect of moisture conditions upon the interpretation of resistivity, absorption and permeability test results is critical. The techniques described below permit quantitative assessments of varying accuracy, and other methods described in Chapter 8 such as infra-red thermography and short wave radar may detect moisture variations on a comparative basis.

7.2.1 Simple methods

One simple approach that is often adopted is to measure the moisture content, by oven drying and weighing, of a small timber insert or mortar prism (191) which has been sealed into a surface-drilled hole. The insert must be left in the sealed hole for sufficient time to reach moisture equilibrium with the surrounding concrete. A chemically based disposable humidity meter is also available from Denmark (192) which is inserted into a surface-drilled hole. Little published evidence concerning its reliability is however currently available.

7.2.2 Neutron moisture gauges

These work on the principle that hydrogen rapidly decreases the energy of 'fast' or high-energy neutrons. Hydrogen is present in water in the concrete, and if the retarded or scattered neutrons resulting from interaction of 'fast' neutrons with a matrix containing hydrogen are counted, the hydrogen, and hence moisture content, can be assessed. The most commonly used sources produce neutrons indirectly by the interaction of α-particles generated by the decay of an X-ray emitting isotope, such as radium, with beryllium.

Although instruments are commercially available that could be used to determine the moisture content of concrete, these are normally calibrated for a sample of semi-infinite volume and uniform moisture content. Since they operate on the basis of measurement of backscatter of the neutrons (see section 8.4.3), results will only relate to a surface zone not exceeding 65 mm to 90 mm deep (according to equipment used). Further sources of inaccuracy in this approach will be moisture gradients near to the surface and the presence of neutron absorbers.

Calibration is not simple and accuracy generally improves with increasing moisture content. Although the equipment is widely used in soils testing, the accuracy possible with the lower moisture levels found in concrete is generally poor.

7.2.3 Electrical methods

As indicated in section 7.1.3.2, in-situ electrical resistivity measurements are influenced by moisture content, but as yet this approach cannot be used to assess moisture other than in very broad classification bands and on a comparative basis.

The change in dielectric properties of concrete with moisture content has, however, been used by a number of investigators as the basis of tests on hardened cement pastes and concretes. This approach is based on measurement of the dielectric constant and dissipation factor. The properties of a capacitor formed by two parallel conductive plates depend upon the character of the separating medium. The dielectric constant is defined as the ratio of capacitances of the same plates when separated by the medium under test, and by a vacuum. When a potential difference is applied to the plates, opposite charges will accumulate, and if the separating medium is ideal these will remain constant and no current will flow. In practice, electron drift will occur and a 'conduction' current flows, and the ratio of this current to the initial charging current is the dissipation factor.

Hammond and Robson (193) reported in 1955, on the basis of limited laboratory studies, that the dielectric constant for pastes was higher than for concretes, and that this constant decreased with age and increasing frequency and increased with water content. Bell et al. (194) have shown that the moisture content of laboratory specimens can be determined to $\pm 0.25\%$ for values less than 6% using a 10 mHz frequency, and Jones (195) has confirmed that high frequencies (10–100 mHz) minimize the influences of dissolved salts and faulty electrode contacts.

Simple hand-held electronic equipment for site usage has recently become available (196) which measures the relative humidity of the air within the concrete using similar principles. The digital meter is connected to an electronic probe sealed into a 60 mm deep surface-drilled 25 mm diameter hole (Figure 7.13). The probe is in the form of a small plastic capacitor in a protective guard and is influenced by the relative humidity of the air in the hole. A range of 20% to 90% RH is quoted, although calibration is based on providing a maximum accuracy at around 75% RH (which is critical for alkali-silica reactions). Accuracy varies with the difference in reading from this value and $\pm 15\%$ of the difference is claimed (e.g. $\pm 3.75\%$ at 50% RH).

This equipment is particularly suited to long-term monitoring situations since the probes may be left in place and readings taken whenever required. Similar probes may also be monitored by automatic recording equipment.

More accurate evaluation of the relative humidity of the air inside a hole drilled into the concrete can be obtained using a chilled mirror hygrometer.

Figure 7.13 Scribe humidity meter.

This device works on the principle of detecting the dew point temperature as condensation forms on a small gold mirror, which can be both heated and cooled from the ambient temperature. The instrument measures relative humidity over the entire range 0–100% with an accuracy of ±1% and a probe developed for use in concrete utilizes an expandable gland (Figure 7.14) to enable a seal within a 26 mm diameter hole drilled into the concrete element.

Studies by Parrott (197) have shown an empirical relationship between the relative humidity of an air void within the concrete and the moisture content of the concrete, but this relationship is not unique and varies with the mix of concrete used. These studies here also demonstrated the high sensitivity of the air permeability of the concrete to the moisture content.

Relative humidity measurements are also reported (198) taken from concrete dust (Figure 7.14) obtained by in-situ drilling. A technique is described (199) where drill dust is collected without excessive evaporation of the moisture and evaluation of the moisture content of the in-situ concrete to an accuracy of ±1% is claimed.

7.2.4 *Microwave absorption*

Microwaves are electromagnetic and have a frequency in the range 10^9–10^{12} Hz. They are absorbed by water at a higher rate than by concrete in which the water may be dispersed, hence measurement of attenuation can provide a method of determining moisture content. The principles of this approach have been described by Browne (200), whilst Boot and Watson (201) have reported the use of transmission methods in concrete.

Figure 7.14 Chilled mirror relative humidity meter (photograph by courtesy of Protimeter Plc).

A microwave beam is generated by a portable radio transmitter and received by a crystal detector connected to an amplifier. Boot and Watson (201) used a frequency of 3×10^9 Hz modulated at 3 kHz by a square wave, with an attenuator tuned to 3 kHz attached to the receiver. The beam is passed through the specimen with the transmitter and receiver located at fixed distances on either side, and the attenuator adjusted to give a null reading. This procedure is repeated with the specimen removed. The difference between the two attenuation readings is due to the specimen, and may be converted to a moisture content with the aid of an empirically developed calibration chart.

Unfortunately, because of the internal scattering and diffraction caused by the heterogeneous nature of concrete, the accuracy of predicted moisture content is low and may be as poor as $\pm 30\%$. Although the method is unlikely to be of much practical value at this level of accuracy, the techniques are still under development and improvements are to be hoped for. The commercially available equipment at present also requires two opposite exposed faces, thus imposing a further restriction on use.

More recent studies (202, 203) of the magnitude of the reflection of an electromagnetic radar wave in the frequency range $0.5–1.0 \times 10^9$ Hz, using an impulse antenna, have also investigated the moisture content of concrete. Access to only one face is necessary and the moisture content measurement is related to the surface zone of concrete. Electromagnetic radiation is also used to detect similar thicknesses, reinforcing bars and voids. A fuller description of this is given in Chapter 8.

Figg (204) has described the use of a microwave moisture meter of 2450 MHz frequency on sections sawn from commercially-produced precast cladding panels. These were constructed from flint and limestone aggregate concretes with an exposed aggregate finish on one face. The tests confirmed a linear relationship between microwave attenuation and percentage of absorbed water, but with a slope which varies with different types of concrete. Prior preparation of a specific calibration chart is therefore necessary to permit an absolute value of moisture content to be obtained in a practical situation. A further limitation to practical use was also indicated by anomalous results obtained during freezing weather.

7.3 Absorption and permeability tests

These properties are particularly important in concrete used for water-retaining structures or watertight basements, as well as being critical for durability. Tests are available for assessing water absorption as well as gas and water permeability. Considerable attention has been paid to this area recently and the topic is covered comprehensively in the Concrete Society Technical Report 31 (205) and by Basheer (206) which both review a wide range of testing techniques and fundamental theory. Only a selection of these test methods is included in this chapter.

The terms permeability and porosity are often used as though their meanings were synonymous. Porous concrete is often permeable and vice-versa. However, the term porosity is a volume property that relates the volume of the pores to the total volume. Unless the pores are interconnected (Figure 7.15) the material will not be permeable. The term permeability normally relates to the ease with which a fluid will pass through a material under a pressure gradient, but is also used to describe capillary, diffusion, adsorption and absorption processes (205).

The mechanism of water transport through an interconnected pore system is schematically shown in Figure 7.16, where a single pore has restrictions on each side. Under dry conditions the principal mechanism is molecular adsorption to the sides of the pores (stage a). Once the pore walls have reached their adsorption limits (stage b), water vapour will diffuse across the pore. Under moisture conditions the restrictions to the pore inlet and outlet become blocked with water (stage c) and further water movement must involve a transfer across this film. At much higher moisture levels (stages d to f) water is transported by liquid flow through the pore.

Ionic diffusion occurs as a result of a concentration gradient rather than a pressure gradient and will occur through the liquid contained in partially or fully saturated pores (stage g).

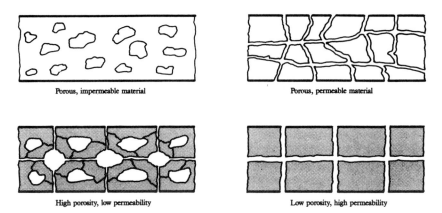

Figure 7.15 Porosity and permeability (based on ref. 205).

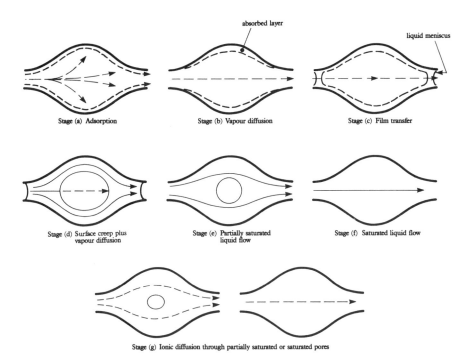

Figure 7.16 Movement of water and ions within concrete pore (based on ref. 205).

The two most widely established methods for in-situ usage are the initial surface absorption test (ISAT) which is detailed in BS 1881: Part 5 (207) [to be replaced by Part 208 in due course] and assesses water absorption, and the modified Figg air permeability method. Several other surface-zone gas and water permeability methods (often similar) are available, but reported experience is limited. These include the 'cover concrete permeability tester' (13) developed by the British Cement Association, which is based on the time for air pressure in a drilled cavity to dissipate and also incorporates relative humidity measurement for incorporation into interpretation. The initial surface absorption method was originally devised for testing precast concrete products, and presents some practical difficulties in use on site. Although the Figg method was developed for site use, and is growing in popularity, reported field experience is not extensive. In all cases emphasis must be on comparative rather than quantitative application, and results only relate to surface-zone properties. It must be recognized, however, that this is the critical region as far as durability performance is concerned.

Laboratory specimens may be tested to measure water flow, and it is possible that suitably prepared samples taken from site may give useful results, although there is no standard procedure available for this approach.

Absorption by an immersed core is also covered by BS 1881: Part 122 (208) but this is again directed towards control testing of precast components. Observation of the rate of penetration of moisture from one wetted surface may be valuable as a comparative test in the laboratory, and some applications of this are outlined below.

Generalized relationships between results of some of the more common test methods are summarized in Table 7.3 (205, 209).

Further, more detailed guidance concerning the application of test results is given by the Concrete Society (205), but the interpretation of site results is hampered by the major influence of existing moisture conditions. Many of these tests are nevertheless used very successfully under laboratory conditions for assessing the characteristics of differing concrete types, constituents, and curing treatments.

Table 7.3 General relationship between permeability and absorption test results on dry concrete (205, 209)

| Method | Concrete permeability/absorption | | |
	Low	Average	High
'Intrinsic permeability' k (m^2)	$<10^{-19}$	10^{-19}–10^{-17}	$>10^{-17}$
ISAT—10 min (ml/m^2/s)	<0.25	0.25–0.50	>0.50
Figg water (s)	>200	100–200	<100
Modified Figg air (s)	>300	100–300	<100
BS Water absorption—			
30 min (%)	<3	3–5	>5
DIN 1048—4 day (mm)	<30	30–60	>60

Note: See ref. 205 for details of calculation of water and gas coefficient of permeability from 'intrinsic permeability' k. Approximate values for water at 20°C $K_w = 9.75 \times 10^6 k$ m/s, air at 20°C $K_g = 6.5 \times 10^5 k$ m/s.

7.3.1   *Initial surface absorption test*

This method is detailed in BS 1881: Part 5 (207), and Levitt (210) has discussed the theory and application of the technique.

7.3.1.1   *Theory.*

Initial surface absorption is defined as the rate of flow of water into concrete per unit area at a stated interval from the start of the test at a constant applied head and temperature. Results will be expressed as $ml/ml^2/s$ at a stated time from the start of test.

When water comes into contact with dry concrete it is absorbed by capillary action, at a rate which is high initially, but decreases as the waterfilled length of capillaries increases. Levitt (210) has shown that this may be described mathematically by the expressions

$$p = at^{-n}$$

where $p =$ initial surface absorption
  $t =$ time from start
  $a =$ a constant
  $n =$ a parameter between 0.3 and 0.7 depending on the degree of silting or flushing mechanisms, but constant for a given specimen.

The standard procedure described by BS 1881 involves a constant head of 200 mm, with readings of initial surface absorption taken at 10 minutes, 30 minutes, 1 hour and 2 hours. Levitt has demonstrated that the factors a and n in the above expression can be evaluated from the 10 minute and 1 hour rates with sufficient accuracy to predict rates at other times. Two-hour readings are seldom used in practice

7.3.1.2   *Equipment and procedure.*

The equipment (Figure 7.17) consists of a cap which may be clamped and sealed onto the concrete surface, together

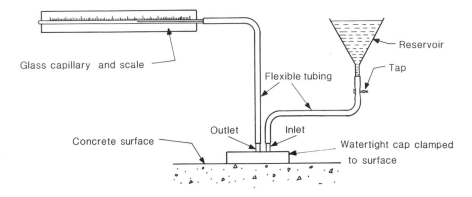

Figure 7.17 Initial surface absorption test.

with an inlet connected to a reservoir and an outlet connected to a capillary tube with a scale. The water contact area must be at least 5000 mm² whilst the reservoir and horizontal capillary must be set at 200 mm ±5 mm above the surface. It is useful if the cap is manufactured from a transparent material to permit visual checking that no air is trapped during a test. The capillary tube should be between 100 mm and 1000 mm with a bore of 0.4–1.0 mm radius which must be determined precisely by measuring the discharge under a 200 mm head. After correcting for temperature effects on viscosity the radius may be calculated from the following simplified formula:

$$r^4 = \frac{L}{20\mu t}$$

where r = radius (mm)
 L = length of capillary (mm)
 t = average time to collect 10 ml of water (s)
 μ = a viscosity factor ranging linearly from 3.73 at 18°C to 4.09 at 22°C.

BS 1881: Part 5 describes a detailed procedure for cleaning the capillary and performing the above measurement.

   The bore of the capillary must be known, to permit calibration of the scale, which is arranged so that the measured movement of water along the capillary during a one-minute period of test will be equal to the value of initial surface absorption at that time. This is achieved by marking the scale in units of 0.01, spaced $(6A/\pi r^2 \times 10^{-4})$ mm apart, where A = water contact area (mm²). At the start of the test, the cap must be fixed to the surface and sealed to provide a watertight assembly. The choice between greased solid rubber gaskets, foamed rubber gaskets or a knife-edge bedded in a sealing material will depend upon the condition of the surface. Before filling, the seal should be tested by gentle blowing, with soap solution applied to the outside of the joint to detect leaks. The reservoir and capillary must be set 200 ± 20 mm above the surface, this being measured from the mid-height of concrete under the cap for non-horizontal surfaces, although it must be possible to lift the end of the capillary to avoid overflow between readings. Timing is started when the reservoir tap is opened and water at 20° ± 2°C allowed to flow into the cap. The capillary should be disconnected from the outlet tube until all air has been expelled. The reservoir head must be maintained, and shortly before a 'measurement time' the capillary adjusted so that it fills with water before fixing horizontally at the same level as the reservoir surface.

   Measurements are made by closing the inlet tap and watching for movement of the capillary. Using a stopwatch, the number of scale units moved in the first five seconds from the start of movement is noted and used to determine the total measurement time. If less than three divisions, measurements should be continued for two minutes, if 3–9 divisions, continue for one minute, or if 10–30 divisions, continue for 30 seconds. The measured reading should then be factored as necessary to give the number of scale units moved in one minute, and hence the value of initial surface absorption. If the movement is more than 30 divisions in five seconds, the result can only be quoted as more than $3.60 \, \text{ml/m}^2/\text{s}$. This procedure should be performed 10 min, 30 min, one hour and two hours after the start of the test, and at least three separate samples tested in this way.

Laboratory samples for testing should be oven-dry, but for large units or samples this is not possible. If the test is to be performed in a laboratory, the concrete must have been in the dry laboratory atmosphere for at least 48 hours. For in-situ testing a minimum dry period of 48 hours is specified, but these requirements are very unlikely to produce comparable moisture conditions within the concrete and this will cause major problems in the interpretation of results.

An in-situ vacuum drying technique has been developed by Dhir *et al.* (211) which overcomes the difficulty of the uncertain moisture content of the in-situ concrete and gives ISAT 10-minute results with a standard deviation similar to that obtained with oven-dried specimens. Several other surface-drying procedures involving the use of heaters have been proposed but, as yet, no standardized procedure has been agreed.

A further development of the ISAT test (212) proposes the use of a guard ring located around the perimeter of the standard cap containing water at the same hydrostatic pressure. Water flowing into the concrete surface from the guard ring will confine the absorption of water from the central cap to a uniaxial flow in the concrete. Using this modification, an improved correlation with the cumulative 2 hour ISAT result and with adsorption tests carried out on 50 mm oven-dried cores was found.

7.3.1.3   Reliability, interpretation and applications.   It has been found that tests on oven-dried specimens give reasonably consistent results, but that in other cases results are less reliable. Particular difficulties have been encountered with in-situ use in achieving a watertight fixing. Levitt (210) has suggested that specific limits could be laid down as an acceptability criterion for various types of construction, but insufficient evidence is yet available. Values of greater than $0.50 \, \text{ml/m}^2/\text{s}$ at 10 minutes for dry concrete have been tentatively suggested as corresponding to a high absorption (see Table 7.3) whilst Dhir, Hewlett and Chan (213) have reported repeatability tests. The test has been found to be very sensitive to changes in quality and to correlate with observed weathering behaviour. The main application is as a quality control test for precast units but application to durability assessment of in-situ concrete is growing. A minimum edge distance of 30 mm is proposed (213). The pressure used is low and although this may relate well to normal surface weather exposure conditions it is of little relevance to general behaviour under higher pressures. The method can be applied to exposed aggregate or profiled surfaces provided that an effective seal can be obtained, but is not suitable for porous or honeycombed concretes.

Dhir *et al.* (214) have also considered use of the 10 minute ISAT result to evaluate resistance of the concrete to chloride diffusion and to carbonation and give guidance for lightweight concrete and for concrete with cement replacements in addition to normal o.p.c. concrete.

7.3.2 *Figg air and water permeability tests*

Figg (215) in 1973 described the development of a test for air and water permeability which involved a hole drilled into the concrete surface. This was commonly known as the *Building Research Establishment Test*. A number of versions of this approach have subsequently been developed in various countries (205) but the most widely accepted procedure is that proposed by Cather, Figg, Marsden and O'Brien (216) on the basis of extensive experience with the method. This is commonly known as the modified Figg method and is described below. Earlier work (215) was based on a smaller hole depth and diameter (30 mm and 5.5 mm respectively) and comparison of results from different investigations should be treated cautiously because of the lack of standardization of hole sizes.

7.3.2.1 Equipment and procedure. A hole of 10 mm diameter is drilled 40 mm deep normal to the concrete surface with a masonry drill. After cleaning, a disc of 3 mm thick polyether foam sheet is pressed 20 mm into the hole and a catalysed liquid silicone rubber is added. This hardens to provide a resilient seal to the small cavity in the concrete, and a gas-and-liquid-tight seal is obtained by a hypodermic needle through this plug.

Air permeability measurements are by means of a hand-operated vacuum pump and digital manometer connected by a three-way tap and plastic tubing to the hypodermic needle as illustrated in Figure 7.18. The pressure is reduced to −55 kPa and the pump then isolated, with the manometer and concrete connected together. The time, in seconds, for air to permeate through the concrete to increase the cavity pressure to −50 kPa is noted and taken as a measure of the air permeability of the concrete.

Water permeabiity is measured at a head of 100 mm with a very fine canula passing through the hypodermic needle to touch the base of the cavity. A two-way connector is used to connect this to a horizontal capillary tube set 100 mm above the base of the cavity, and a syringe. Water is injected by syringe to displace all air, and after one minute the syringe is isolated with the water meniscus in the capillary in a suitable position. The time, in seconds, for the meniscus to move 50 mm is taken as a measure of water permeability of the concrete.

An automated system (217), the 'poroscope' utilizes a preformed silicone plug (Figure 7.19) to avoid delays in waiting for sealant to set and can be used for both air and water permeability Figg tests. Results obtained are

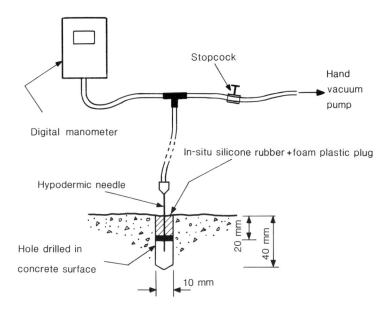

Figure 7.18 Modified Figg air permeability test (based on ref. 216).

similar to those using a liquid silicone sealant and manual control. The use of a preformed plug also enables tests to be carried out on a slab soffit, which was not previously possible using a liquid sealant (218).

7.3.2.2 Reliability, limitations and applications. Using this method, the relationships between air pressure and time, and meniscus movement and time, were both found to be nearly linear. The test criteria of hole depth, plug thickness, pressure and test time have been investigated. Results have been obtained on laboratory concrete which show that both air and water permeability measured in this way correlate well with water/cement ratio, strength and ultrasonic pulse velocity. Aggregate characteristics have a profound effect on results, limiting the potential usage to comparative testing, but variations of drilling and plugging of the test hole are less significant. As with the initial surface absorption method, the moisture condition of the concrete will considerably influence the results. This seriously restricts in-situ usage, but a general classification for dry concrete is given in Table 7.3. Values should be treated with caution because of variations in hole dimensions used by various investigators. It has also been suggested that use of an even larger hole will improve repeatability (213).

The principal use of this method, which involves cheap and simple techniques, is as an alternative to the initial surface absorption method for quality control checking in relation to durability. It is usually the outer 50 mm

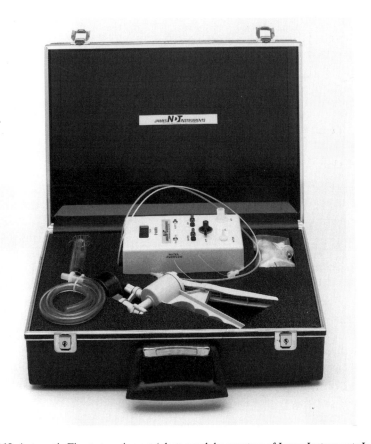

Figure 7.19 Automatic Figg test equipment (photograph by courtesy of James Instruments Inc).

of concrete which is most important when considering durability, since this protects the reinforcement from corrosion. The method tests this region, and has the advantage of not being influenced by very localized surface effects such as carbonation of the outer few millimetres of the concrete.

7.3.3 Combined ISAT and Figg methods

Dhir, Hewlett and Chan (213) have proposed that the shortcomings of the ISAT and Figg water permeability methods can be overcome by combining these in the form of the Covercrete absorption test (CAT). It is intended that the test be performed on 100 mm cores taken from the structure which are oven dried to overcome moisture effects. A 13 mm diameter 50 mm deep hole is drilled and ISAT equipment with a 200 mm head of water is used but with a 13 mm internal diameter cap and the inlet tube located inside the sealed hole. A coefficient of variation of about 8% was obtained with this approach which is significantly better than for the Figg water method.

7.3.4   Hansen gas permeability test

This Danish method is described in detail by Hansen, Ottosen and Petersen (219). An 18 mm hole is drilled at a shallow angle below the concrete surface. The pressure rise within this hole is monitored by a sensor inserted in the hole as compressed air pressure is applied to a sealed surface region. The equipment is commercially available, although published experience is limited at present.

7.3.5   'Autoclam' permeability system

The hydrostatic pressure used in the ISAT is low and there are circumstances related to severe exposure conditions or testing of surface coatings where a higher pressure in-situ test is needed. The Autoclam is a commercially available system developed at Queen's University, Belfast (220) which is similar in principle to the ISAT, but which uses a hydrostatic pressure of 1.5 bar to measure in-situ water permeability (Figure 7.20). The equipment can alternatively be used to measure a low-pressure, 0.01 bar water sorptivity. In addition a measure of air permeability, through decay of air pressure applied to the concrete surface, can also be measured. All three tests are carried out by clamping the Autoclam to a 50 mm metal ring which is glued to the concrete surface. The low-pressure water sorptivity test can be carried out at the same location as the air permeability test, but it is recommended that the high-pressure water permeability test is done at a different test location.

Once each test is commenced, the control of the test and monitoring of results is fully automatic. Each test takes about 15 minutes and all tests are

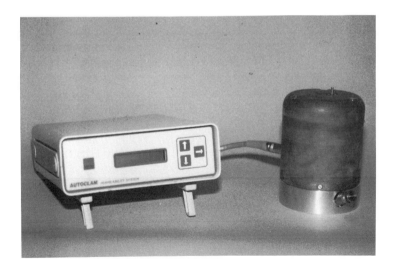

Figure 7.20 AUTOCLAM air and water permeability instrumentation.

sensitive to the moisture condition of the in-situ concrete. On laboratory cubes, the influence of the internal moisture in the concrete can be eliminated by oven drying to a constant weight. For in-situ concrete, a method of surface-drying for 20 minutes using a propane heater followed by cooling for 1 hour was found to give reliable results unaffected by the original moisture condition of the concrete. The Autoclam tests were found to correlate well with other permeability and sorptivity tests and can be used to rank different concretes into poor, medium and high durability categories. The high-pressure permeability test is thought to test concrete up to 40 mm below the surface and the results have the potential to be used to evaluate the intrinsic water permeability of the concrete (221).

7.3.6 *Flow tests*
High-pressure permeability tests hve been used in the laboratory for many years and consist of the measurement of the steady flow of water through a concrete specimen under a specified head. The equipment is usually simple and an arrangement of the form shown in Figure 7.21 may be suitable. The specimen is sealed in a permeameter so that air or water under pressure can

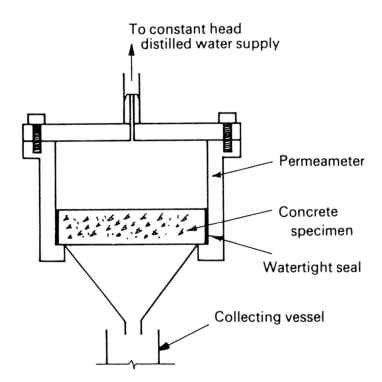

Figure 7.21 Typical permeability equipment.

be applied to one face and the amount of fluid that permeates and emerges from the other face is measured. Leakage must be avoided and a constant head must be maintained whilst measuring the fluid passing through the sample; precautions must also be taken to avoid evaporation losses from the reservoir when water is being used.

Van der Meulen and Van Dijk (222) have described an approach of this type and paid particular attention to the method of sealing the specimen. Whilst this may be reasonably easy for laboratory cast specimens, if the concrete has been obtained from an in-situ location, by coring, the surface will be irregular. It is suggested that a specimen of this type can be cast into an epoxy ring which has an accurate outer surface and can then be fitted, with the aid of neoprene seals, into a brass ring mounted in the permeameter pot.

The observed flow will be of the form shown in Figure 7.22, and the test is usually continued to enable a steady state to be reached. The results from samples of similar dimensions tested under similar heads may be used comparatively to assess the characteristics of the interior of a body of concrete, and this approach is most likely to find applications in the field of water-containing structures.

7.3.7 BS absorption test

BS 1881: Part 122 (208) describes a standard absorption test to be carried out on 75 mm diameter (± 3 mm) cored specimens. The test is intended as a durability quality control check and the specified test age is 28–32 days. The apparatus required is simple, consisting only of a balance, an airtight vessel, a container of water and an oven, and the test is also straightforward to perform.

The procedure involves drying the measured core specimen, which should be 75 mm long if the member thickness is greater than 150 mm, in an oven

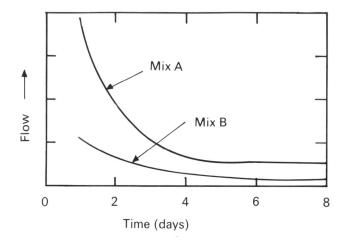

Figure 7.22 Typical flow results.

Table 7.4 Water absorption correction factors (based on ref. 208)

Length in mm of 75 mm dia. specimen	Absorption correction factor
35	0.73
50	0.86
75	1.00
100	1.09
125	1.16
150	1.20

at 105°C $\pm$ 5°C for 72 $\pm$ 2 hr. The specimen is cooled for 24 $\pm \frac{1}{2}$ hr in an airtight vessel, weighed and immersed horizontally in the tank of water at 20 $\pm$ 1°C with 25 $\pm$ 5 mm water cover over the top surface. It is immersed for 30 $\pm \frac{1}{2}$ min, removed, shaken and dried quickly with a cloth to remove free surface water before weighing again.

The absorption is calculated as the increase in weight expressed as a percentage of the weight of the dry specimen, and a length correction factor is applied if necessary. Some typical values of this factor which yield the equivalent absorption of a 75 mm long core are given in Table 7.4. The results are expressed to the nearest 0.1% and it is recommended that at least three samples are tested and the result averaged. Results are classified in general terms in Table 7.3.

The reliability of the method may be influenced by the effects of coring, and the measurements do not relate to the exposed concrete surface. Levitt (210) has also suggested that air trapped at the centre of the specimen affects readings after more than 10 minutes' immersion. He suggests that results taken at this time would be more reliable and may be related to initial surface absorption values after 10 minutes, but the method is less sensitive to concrete quality and the testing accuracy is also lower. The surface drying prior to weighing is particularly likely to influence results. If results are to be compared with a specified limit, or with other values, the age at test is particularly important. An appreciably higher value would be expected at ages under 28 days, and lower values at greater ages, due to a number of factors including the rate of hydration of the cement. This will therefore be a characteristic of the particular mix used and limits the general application of the method other than as a quality control test.

7.3.8 DIN 1048 absorption test (223)

This test is usually performed on purpose-made 200 mm square laboratory blocks, although successful use on 150 mm diameter cores has been reported (224). Specified water pressures are applied to a sealed 100 mm diameter area of the cast face over a period of 96 hours. The specimen is then split to measure the maximum depth of water penetration, and the results of three such specimens averaged.

7.3.9 'Sorptivity' test

The mechanisms of water penetration into unsaturated concrete have been analysed by Kelham (225) who proposes a technique for measuring the critical material constant known as the sorptivity. The test is based on measuring the weight change of a submerged 50 mm thick 150 mm diameter cylindrical sample as water is absorbed through one of the flat faces. The opposite face is exposed to the atmosphere whilst the curved surface is sealed against water penetration. The sorptivity is derived from the slope of a plot of sample weight against the square root of time. The effective porosity can also be obtained by allowing the sample to become saturated.

Although reported results are for laboratory cast specimens, slices of cores could be used after suitable moisture conditioning.

7.3.10 Capillary rise test

A porosity test to assess the quality of high alumina cement concrete has been described by the Institution of Structural Engineers (226). A piece of concrete at least 50 mm long is sawn flat on one face and placed with this surface on a block of felt or blotting paper 10 mm thick. This is placed in a dish containing water to a level below the top surface of the felt, and the dish covered by an open-mouthed bell jar to prevent evaporation.

Capillary rise is measured by visual observation of colour change, and readings up to 8 hours and at 24 hours are suggested. The rise of water in the specimen at these times can be related to percentage conversion of the concrete, with values of 15–20 mm at 24 hours quoted for unconverted concrete. This represents good quality, well compacted concrete; 25–35 mm may be expected for similar concrete if unfavourably converted. A higher value may indicate serious deterioration. Whilst this method is described in relation to one specific application, it may have other worthwhile applications of a comparative nature both in the laboratory and on site.

7.3.11 Vacuum test

A vacuum test has been developed in the USA (227) which measures the air permeability of concrete by studying the decay in the vacuum as air passes from the cover zone into a cap sealed to the surface. To date, this test has seen very little usage elsewhere.

7.4 Tests for alkali-aggregate reaction

Reactions between the aggregates and alkaline matrix pore fluids are common in some parts of the world, and have been the subject of widespread research. A wide variety of such reactions exist which generally cause internal expansion in the presence of moisture due to the formation of hygroscopic gel at the aggregate/matrix interface and within aggregate particles. In the UK the most common form experienced is alkali-silica reaction.

Several tests exist to aid assessment of potential reactivity of materials, which are outside the scope of this book. In-situ tests related to diagnosis, and assessment of existing and future damage are not well developed although ultrasonic pulse velocity and pulse-echo techniques may offer potential. Assessment of existing structures is primarily based on cores which may be subjected to expansion tests, alkali-immersion tests, cement content and petrographic (sawn surface and thin section) analysis (see Chapter 9). Visual identification of crack characteristics as well as a knowledge of moisture content will also be important aids to diagnosis, which is considered in detail by a recent British Cement Association report (23).

Expansion tests basically involve strain measurements on cores with at least 4 Demec gauge spans along their length. The reaction may be accelerated for diagnosis purposes by a 38°C, 100% RH environment (23), although a number of variations on this are possible to obtain an indication of likely future performance. A 'stiffness damage' test on cores has also been described in section 5.1.3.6, and a technique of staining the alkali reaction gel with an ultra-violet fluorescent dye to facilitate visual identification is discussed in section 9.11.1.

7.5   Tests for freeze-thaw resistance

These usually take the form of petrographic analysis of a polished surface of a sample removed from the structure to determine entrained air content. ASTM C457 (228) covers this type of testing which is considered further in section 9.11. Laboratory temperature cycling of samples may also be used.

7.6   Abrasion resistance testing

Abrasion resistance is likely to be of critical importance for floors of industrial premises such as factories or warehouses where disputes involving very high repair or replacement costs may often arise. The pore structure of the surface zone has been found to be the principal determining factor by Sadegzadeh, Page and Kettle (229) using an accelerated wear apparatus. This consists of a rotating steel plate carrying three case-hardened steel wheels which wear a 20 mm wide groove with 205 mm inside diameter in the concrete surface. Abrasion resistance is assessed by measurement of the depth of this groove after a 15 minute standardized test period. Field studies using this equipment and rebound hammer tests are reported by Kettle and Sadegzadeh (230). These cover a range of practical situations and surface finishing techniques, and it is concluded that the relationship between rebound number and abrasion resistance is more complex than previously proposed by Chaplin

(52). Results from the accelerated-wear equipment correlated well with observed deterioration, and the classification in Table 7.5 is proposed for slabs in a medium industrial environment. Kettle and Sadegzadeh (231) have also shown that the ISAT is very sensitive to factors influencing abrasion resistance and may well provide an indirect non-destructive approach.

Table 7.5 Classification of concrete floor slabs in medium industrial environment (based on ref. 230)

Quality of slab	Abrasion depth (mm)
Good	<0.2
Normal	$0.2–0.4$
Poor	>0.4

8 Performance and integrity tests

The tests described in this chapter are wide ranging and measure a variety of different properties related to performance and integrity which have not been considered elsewhere in the book. Many of the durability tests described in Chapter 7 and chemical tests in Chapter 9 are also related to performance, and load testing has been covered in Chapter 6. Integrity tests are generally concerned with the location and development of voids and cracks which are not visible at the surface but may influence structural or durability performance.

8.1 Infrared thermography

The use of this technique, in which infrared photographs are taken during the cooling of a heated structure, has been described by Manning and Holt (232). Infrared thermography offers many potential advantages over physical methods for the detection of delamination of bridge decks which are discussed in section 8.3. The detection of laminations or voids by infrared thermography is based on the difference in surface temperature between sound and unsound concrete under certain atmospheric conditions. The concrete surface temperature changes throughout the day due to heating by sunlight and cooling at night, this being particularly marked in summer. During the hottest part of the day the concrete temperature decreases with depth below the surface, whereas at night the situation is reversed. Delamination will affect these temperature gradients, and hence surface temperatures, which may be compared by infrared measurements. Because of the small temperature differentials involved, the results cannot be directly recorded by infrared film, and the image from the camera must be displayed on a cathode ray tube. The temperature of the surface is indicated by a range of greys, although thermal contours can be automatically superimposed, and colour monitors are also available. This image is normally recorded on film to provide a hard copy.

Early trials have been performed with a camera held by an operator standing on the deck (232), but the limited field of view and oblique alignment render this impracticable. Greater success was obtained by using a bucket truck to scan the deck from a height of up to 20 m, provided that the surface temperature differentials were greater than 2°C. Airborne testing avoided the necessity for lane closures and reduced the difficulties of piecing together photographs to form a composite picture (the equipment was mounted on a helicopter and trial flights were made at heights up to 400 m) but the images obtained were often poor.

It was found that surface temperature differentials existed in delaminated decks at most times, but were greatest in summer, peaking in mid-afternoon. Attenuation of the reflected heat by the atmosphere, however, poses a major problem, and this is increased by wind. Moisture on the surface was also found to mask surface temperature differences. While the potential advantages of airborne operation are considerable despite the high cost, there remain problems to be resolved before this technique can be considered fully reliable and economical.

Holt and Eales (233) have also described the successful use of thermography to evaluate defects in highway pavements with an infrared scanner and coupled real-life video scanner mounted on a 5 m high mast attached to a van. This is driven at up to 15 mph and images are matched by computer. Procedures for thermography in the investigation of bridge deck delamination are given in ASTM D4788 (234).

Other applications of infrared thermography to structural investigations involve comparative assessment of concrete moisture conditions, which will influence thermal gradients, as well as location of hidden voids or ducts (235). Techniques are also available to detect reinforcing bars which are heated by electrical induction. More recent development of 12 bit equipment (237) has improved the sensitivity to temperature differences to within $\pm 0.1°C$ (Figure 8.1) and enabled high definition imaging and accurate temperature measurement. Studies of mosaic cladding to multi-storey buildings have been conducted to detect delamination and defective areas (238). The location of 'hot spots' during daytime or 'cold spots' at night are used to identify where remediation is required. The smallest detectable defect is reported to be 200 mm × 200 mm. Studies of the insulation properties of buildings (239) using infrared imaging can be used to reduce heat losses at hot spots by identifying missing thermal insulation.

8.2   Radar

Over the past decade there has been an increasing usage of sub-surface impulse radar to investigate civil engineering problems and, in particular, concrete structures (240). Electromagnetic waves, typically in the frequncy range 500 MHz to 1 GHz, will propagate through solids, with the speed and attenuation of the signal influenced by the electrical properties of the solid materials. The dominant physical properties are the electrical permittivity which determines the signal velocity, and the electrical conductivity which determines the signal attenuation (241). Reflections and refractions of the radar wave will occur at interfaces between different materials and the signal returning to the surface antenna can be interpreted to provide an evaluation of the properties and geometry of sub-surface features (Figure 8.2).

Figure 8.1 Infrared thermography apparatus.

8.2.1 *Radar systems*

There are three fundamentally different approaches to using radar to investigate concrete structures.

(i) Frequency modulation: in which the frequency of the transmitted radar signal is continuously swept between pre-defined limits. The return signal is mixed with the currently transmitted signal to give a difference frequency, depending upon the time delay and hence depth of the reflective interface. This system has seen limited use to date on relatively thin walls (242).

(ii) Synthetic pulse radar: in which the frequency of the transmitted radar signal is varied over a series of discontinuous steps. The amplitude and phase of the return signal is analysed and a 'time domain' synthetic pulse is produced. This approach has been used to some extent in the

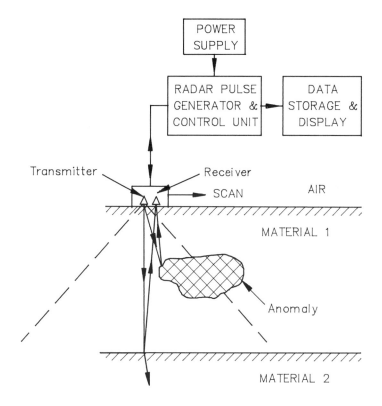

Figure 8.2 Investigation of sub-surface anomaly using radar.

field (243) and also in laboratory transmission line studies (244) to determine the electrical properties of concrete at different radar frequencies.

(iii) Impulse radar: in which a series of discrete sinusoidal pulses within a specified broad frequency band are transmitted into the concrete, typically with a repetition rate of 50 kHz. The transmitted signal is often found to comprise three peaks (Figure 8.3), with a well-defined nominal centre frequency.

Impulse radar systems have gained the greatest acceptance for field use and most commercially obtainable systems are of this type, e.g. as shown in Figure 8.4. The power output of the transmitted radar signal is very low and no special safety precautions are needed. However, in the UK a Department of Trade and Industry Radiocommunications Agency licence is required to permit use of investigative radar equipment.

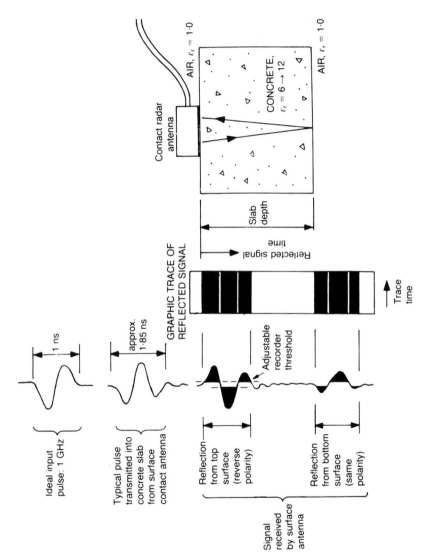

Figure 8.3 Radar reflections from a concrete slab (based on ref. 241).

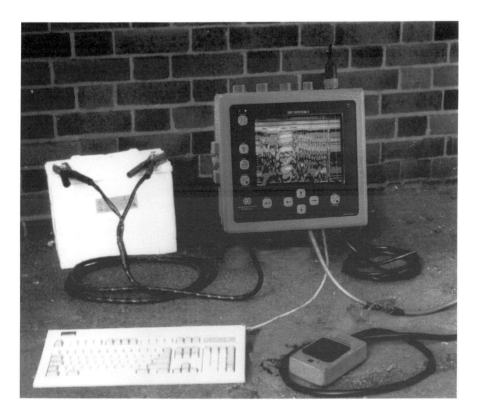

Figure 8.4 Digital impulse radar system.

8.2.1.1 Radar equipment. Impulse radar equipment comprises a pulse generator connected to a transmitting antenna. This is commonly of a dipole 'bow-tie' configuration, which is held in contact with the concrete and produces a diverging beam with a degree of spatial polarization. A centre frequency antenna of 1 GHz is often used for investigation of relatively small concrete elements (up to 500 mm thick), whilst a 500 MHz antenna may be more appropriate for deeper investigations (245). A reduction in antenna frequency results in less signal attenuation and hence a deeper penetration capability, but has the disadvantages of a poorer resolution of detail and the need for an antenna which is physically quite large and cumbersome.

An alternative to using surface-contact antennae is to use a focused beam horn antenna with an air gap of about 300 mm between the horn and the concrete surface. These systems have been used in the USA and Canada (202, 246) to survey bridge decks from a vehicle moving at speeds up to 50 km/h, principally to detect corrosion induced delamination of the reinforced concrete slab. This technique is now well established and operational details are provided in ASTM D4748 (247).

In the UK, specialized antennae have been developed by British Gas (248) and ERA (249), whilst a complete range of commercial general purpose equipment is principally available from GSSI (250) and Sensors and Software (251). A hand-held radar scanner is also under development at the time of writing by Zircon Corp. (252).

8.2.2 *Structural applications and limitations*

In addition to the assessment of concrete bridge decks described in section 8.2.1, radar has been used to detect a variety of features buried within concrete, ranging from reinforcing bars and voids (253) to murder victims (254). The range of principal reported structural applications is summarized in Table 8.1.

Interpretation of radar results to identify and evaluate the dimensions of sub-surface features is not always straightforward. The radar 'picture' obtained often does not resemble the form of the embedded features. Circular reflective section such as metal pipes or reinforcing bars, for example, will present a complex hyperbolic pattern (Figure 8.5a) due to the diverging nature of the beam. The use of signal processing can simplify the image (Figure 8.5b), but there are still three primary hyperbolae, caused by the three peaks of the input signal, which result from a direct reflection off the circular feature as the antenna is scanned across the surface. Below this are three secondary hyperbolae, resulting from a double reflection between the embedded feature and the surface of the concrete. Evaluating the depth of a feature of interest necessitates a foreknowledge of the speed at which radar waves will travel through concrete. This is principally determined by the relative permittivity of the concrete, which in turn is determined predominantly by the moisture

Table 8.1 Structural applications of radar (based on ref. 241)

Reliability		
Greatest→		Least
Determine major construction features		
Assess element thickness		
Locate reinforcing bars		
Locate moisture		
Locate voids/honeycombing/cracking		
Locate chlorides		
Size reinforcing bars		
Size voids		
Estimate chloride concentrations		
Locate reinforcement corrosion		

(a)

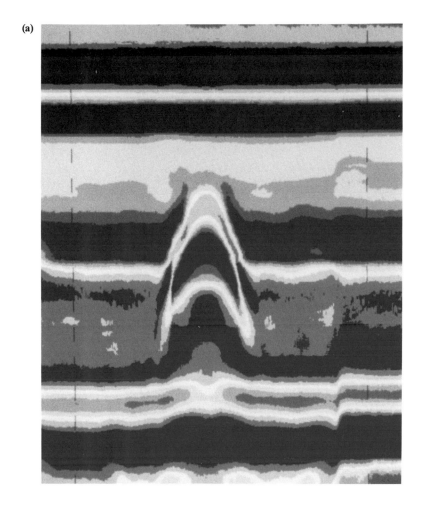

Figure 8.5 Radar scan over steel reinforcing bar. (a) original result; (b) processed result.

content of the concrete. A value of the relative permittivity must either be assumed or the concrete calibrated by localized drilling or coring.

Typical relative permittivity values for concrete range between 5 (oven-dry concrete) and 12 (wet concrete). The propagation of radar through any medium is is governed by complex mathematical expressions. However, for most civil engineering situations using non-magnetic concrete constituents the velocity of the radar signal can be expressed by the simplified equation

$$v = \frac{c}{\varepsilon_r^{1/2}} \text{ m/sec}$$

(b)

Figure 8.5 Continued.

where c = speed of light $(3 \times 10^8 \, \text{m/sec})$
  ε_r = relative permittivity of material.

If the properties of materials are known precisely it may be possible to make depth estimates to within about $\pm 5 \, \text{mm}$, but in practice uncertainties and concrete variability are likely to increase this range significantly.

Because of the difficulties in interpreting radar results, surveys are normally conducted by testing specialists who rely on practical experience, have a knowledge of the fundamental principles involved and have a sense of realism concerning the likely limitations in a given practical situation. Experience has shown (255) that features such as voids can be particularly difficult to detect if located very deep or beneath a layer of closely spaced steel reinforcing bars. The use of neural network 'artificial intelligence' has been used (256) to facilitate the interpretation of radar results and these preliminary studies are encouraging.

Radar reflects most strongly off metallic objects or off an interface between two materials with very different relative permittivities. Thus an air-filled void in dry concrete ($\varepsilon_{r, air} = 1$; $\varepsilon_{r, air dried concrete} = 6$) can be quite difficult to detect, especially if the void is small. Fifty millimetres has been suggested as a practical lower limit to size (253). Conversely, the same void if filled with water ($\varepsilon_{r, water} = 81$; $\varepsilon_{r, wet concrere} = 12$) may be much easier to detect.

Moisture in the concrete will not only influence the speed of the radar signal and the strength of interfacial reflections, but in additon a high conductivity will also cause a greater degree of signal attenuation, resulting in a reduction in the maximum depth of penetration possible. Using a 1 GHz antenna, typical practical penetration limits are around 500 mm for dry concrete and 300 mm for water-saturated concrete. If the concrete is saturated with salt water, the limit of signal penetration is likely to be even smaller.

8.3   Dynamic response testing

Testing ranges from simple surface tapping to investigation of the response of large structures to applied dynamic loads. Simple methods are useful for assessing localized integrity, such as delamination, whilst more complex methods are commonly used for pile integrity testing, determination of member thicknesses, and examination of stiffnesses of members affected by cracking or other deterioration.

8.3.1   *Simple 'non-instrumented' approaches*

Experience has shown that the human ear used in conjunction with surface tapping is the most efficient and economical method of determining major delamination. Delamination may be caused in bridge decks (232) by corrosion expansion of the top layer of reinforcement causing a fracture plane at that level, and is especially likely where cover is low. In order to carry out repairs, the deterioration must be detected and its extent determined. Methods currently in routine use are based on the characteristic dull sound when the deck surface is struck over a delamination. Hammers, steel bars and chains are sometimes used to strike the deck and the sound produced is assessed subjectively by the operator. The operation is performed over a grid to 'map out' the deck. These manual impact techniques have many disadvantages, and 'chain-dragging' provides a quicker and more efficient approach. A single heavy chain with 50 mm links, weighing 2.2 kg/m, has been found to be the optimum, and a 2 m length is passed over the surface with a whip-like action. Sound differences may be detected by the operator and a delamination profile marked. Procedures for using sounding techniques for locating localized delamination in bridge decks are given in ASTM D4580 (257).

This approach, however, relies upon a subjective judgement by the operator to differentiate between sound and unsound regions and results cannot readily be quantified.

8.3.2 Pulse-echo techniques

Various attempts have been made to monitor electronically the response to a blow applied to the concrete surface, and have led to pulse-echo techniques which are now established in a variety of forms. The simplest version commercially available is the *instrumented delamination device* (IDD) which measures the amplitude of the reflected shock waves caused by a surface hammer blow as illustrated in Figure 8.6. If the path of the waves is short a high reading will be obtained, and the measured value will decrease as the path length increases. The readings can only be used comparatively to detect changes in path length due either to variation in member thickness or major internal crack planes.

Standardization of the impact has been one of the major obstacles to quantitative assessment. The associated equipment for processing the signal obtained from the receiving transducer is complex and has been developed for application to the location of voids, ducts, honeycombing and other 'flaws'. Success has been achieved in the laboratory but few reports are yet available of site usage. A 'thickness gauge' is available in the UK.

8.3.2.1 Time domain measurement. The pulse-echo technique is most widely developed for the testing of concrete piles, and is based on the analysis of

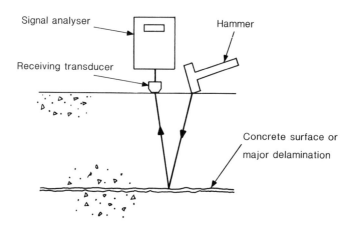

Figure 8.6 Instrumented delamination device.

reflected pulse traces to detect defects or varying soil support conditions. A single blow is applied by hammer to the pile top and the subsequent movements, including reflected shock waves, are detected by a hand-held accelerometer. This is connected to a signal processor which integrates the signal, and a trace of vertical displacement against time can be displayed on an oscilloscope. The basic equipment and a typical trace are shown in Figure 8.7.

The trace may be used for reliable assessment of pile length and to detect any defects or lack of uniformity of the pile, which will show as distortions to the trace. These may be interpreted by a specialist to determine their type and position. The equipment can record and store several hammer blows to check signal consistency, and the oscilloscope traces may be recorded by an instant camera attachment. The method is simple and cheap to use, but cannot determine the cross-sectional area of a pile or its bearing capacity.

This method is most suitable for permanently cased piles or for single lengths of precast piles. The maximum length which can be tested depends upon the ability to produce well defined peaks due to reflections from the toe. Considerable damping will be provided by the surrounding soil and an approximate limiting length/diameter of 30 is often assumed for piles in a medium clay. Defects will cause intermediate but lower peaks although it is difficult to differentiate between bulbs and neckings.

A wide range of systems is commercially available. One form (258) involves a 6 kg hammer falling through a height of one metre to provide the blow, and permits results to be presented on a flat bed plotter. Other systems

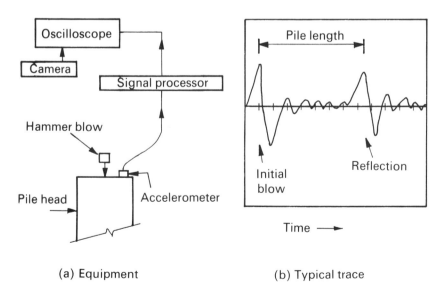

(a) Equipment   (b) Typical trace

Figure 8.7 Pulse-echo method; time domain measurement.

involve various electronic refinements of the signal and trace presentation formats. Practical application of the method has been described (259) with a rate of 50–100 piles a day claimed, and the approach has gained widespread acceptance for quality control and integrity testing of piles.

Time domain measurements have also been used to detect voids in concrete slabs (55) using ultrasonic pulse-echo techniques. The presence of an air void is seen as a faster return of the reflection signal (Figure 8.8).

8.3.2.2   Frequency domain measurement.   Whilst the direct measurement of pulse echo in the time domain is appropriate for relatively large structural elements, such as pile foundations, measurements of thin concrete elements such as slabs or walls result in short-duration, multiple echoes that are time-consuming to measure and can be difficult to interpret. A quicker and simpler approach is to carry out a frequency analysis of the reflection signal

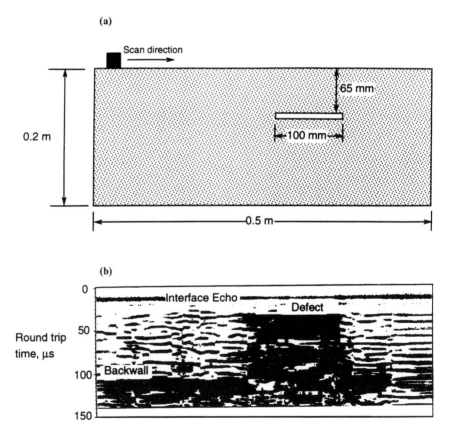

Figure 8.8 Imagining of defect in concrete using ultrasonic pulse-echo technique. (a) Cross section of test specimen; (b) B-scan. Based on ref. 55.

(260). The time domain signal is transformed into the frequency domain using the fast Fourier transform technique. Features such as voids or delamination can be detected by a shift in the amplitude of the higher frequency components of the return signal (Figure 8.9).

The presence of a layer of reinforcing bars will result in a high-frequency signal component, resulting from a reflection from the upper surface of the steel bars (261). By reducing the duration of the input impulse the effect of these bars can be enhanced. By increasing the duration of the impulse the bars can be made 'invisible', thus facilitating the detection of voids beneath the reinforcement. The use of a long-duration impulse does however limit the resolution of voids detectable to sizes greater than the size of the bar diameter.

Impact-echo methods can also be used to detect a loss of contact between the upper surface of a reinforcing bar and the surrounding concrete, but it is difficult to detect a loss of contact on the underside of a bar (261).

A commercial impact-echo package, the DOCter, is now available

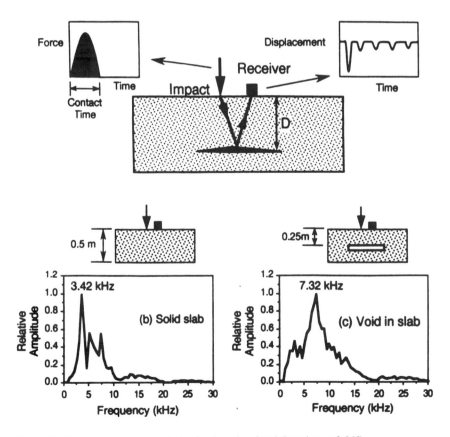

Figure 8.9 Frequency domain analysis of pulse-echo signal (based on ref. 265).

Figure 8.10 DOCter impact-echo apparatus.

(Figure 8.10) and is suitable for use in the field (262). It has been used in the USA to monitor the thickness and integrity of concrete pavements and walls (263) and also for detecting cracking in beams (264). Using an impact duration of 20–80 μs, sections up to 1 m thick have been studied. Studies on the use of impact-echo techniques to detect air voids resulting from inadequate grouting of steel prestressing ducts are promising (265) and recent studies (266) have investigated the use of neural network 'artificial intelligence' to facilitate the interpretation of complex frequency domain results.

8.3.3 *Surface wave techniques*
The use of fast Fourier transformation of impulse responses has also been used for spectral analysis of surface waves (SASW) that are transmitted laterally through a medium (267). A pair of surface transducers are positioned adjacent to the impulse location (Figure 8.11) and are used to measure variations in the surface wave velocity at different frequencies. The results are then further analysed to determine both the thickness and elastic modulus of the material and of underlying layers. This technique has the potential to detect interlayered good and poor quality concrete, but the complex signal processing required has resulted in infrequent usage in the field. Efforts are now being made to automate the analysis of results (268).

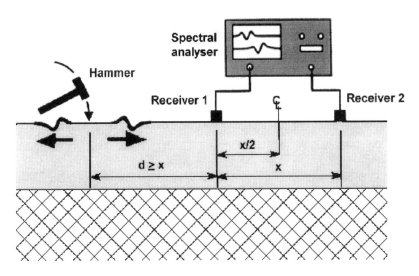

Figure 8.11 Spectral analysis of surface waves.

8.3.4 *Testing large-scale structures*

Dynamic response testing of entire structures may similarly involve hammer impacts or application of vibrating loads. In either case the response is recorded by carefully located accelerometers and complex signal processing equipment is required. Maguire and Severn (269) have described the application of hammer blow techniques to a range of structures including chimneys and bridge beams. Vibration methods have been reported by a number of authors, including Williams (270) and are commercially available.

Model analysis and dynamic stiffness approaches are under development in the UK but are not yet widely used on site.

The aim of testing will usually be to monitor stiffness changes due to cracking, deterioration or repair. The 'dynamic signature' of a structure may be obtained and compared at various time intervals to monitor changes. This is particularly valuable when monitoring the effects of a suspected overload or deterioration with time, and delamination of screeds or toppings from slabs may be detected. Comparisons may in some cases also be possible with theoretically calculated frequencies of vibration and member stiffness.

A range of vibration and 'shock' methods is also regularly used for testing piles. The basic principles have been described by Stain (271) and permit estimates with varying degrees of confidence of dynamic pile head stiffness, cross-sectional area and limiting stiffness values of end bearing piles. As with pulse-echo techniques, the principal advantage is speed of test in comparison with traditional static methods.

8.4   Radiography and radiometry

Radioactive methods have developed steadily over recent years, and although generally expensive and more appropriate to laboratory conditions, their field applications are increasing in number. There are three basic methods currently in use for testing concrete: X-ray radiography, γ-ray radiography, and γ-ray radiometry.

The radiographic methods consist essentially of a 'photograph' taken through a specimen to reveal a picture of the interior, whereas radiometry involves the use of a concentrated source and a detector to pick up and measure the received emissions at a localized point on the member.

X-rays and γ-rays are both at the high energy end of the electro-magnetic spectrum and will penetrate matter, but undergo attenuation according to its nature and thickness. The energies of these rays are expressed in electron volts, and the sources available for concrete testing will typically be in the range 30–125 keV for X-rays and 0.3–1.3 MeV for γ-rays. The attenuation of radiation passing through matter is exponential and may be expressed (62) in simplified terms as

$$I = I_0 e^{-\mu m}$$

where I = emergent intensity of radiation
$\quad I_0$ = incident intensity
$\quad \mu$ = mass absorption coefficient
$\quad m$ = mass per unit area of material traversed.

In the X-ray energy range, attenuation is dependent on both the atomic number and density of the material, whilst for the γ-ray range, density is the principal factor. The cost and immobility of X-ray equipment, which requires high voltages, has been a major limitation to the development of field usage although the method is of considerable value in the laboratory. Sources of γ-rays are more easily portable, and so this has become the principal radioactive method for on-site use.

BS 1881: Part 205 (272) provides guidance on radiographic work, including suitable sources of radiation, safety precautions and testing procedures. Forrester (273) has described the techniques in detail.

8.4.1   *X-radiography*

Laboratory applications of this approach have been principally aimed at the study of the internal structure of concrete, and are summarized by Malhotra (50). Studies have included the arrangement of aggregate particles, including their spacing, and paste film thicknesses, three-dimensional observation of air voids, segregation, and the presence of cracks. Although some field applications were reported in the 1950s, little attention seems to have been paid to the use of X-rays in the field since that time, apart from the use of lorry-mounted high energy 8 MeV Linac equipment capable of penetrating up to 1600 mm of concrete (274).

Concern over corrosion of post-tensioned prestressing strands as a result of inadequate grouting has led to a recent focus of interest in systems capable of investigating the interior of steel prestressing ducts of bridge beams. The Scorpion II system, developed in France, comprises a miniature 4 MeV linear accelerator which produces a beam of high-powered X-rays (275). A lorry-mounted system is used to investigate the integrity of post-tensioned bridge beams by tracking the accelerator on one side of a beam web and locating a detector on the other side. Results are obtained as real-time video display. However, the risk of backscatter radiation reflected back from the concrete surface has necessitated a 250 m exclusion zone below bridges to safeguard the health of pedestrians and boatmen.

8.4.2 *Gamma radiography*

The use of γ-rays to provide a 'photograph' of the interior of a concrete member has gained considerable acceptance. It is especially valuable for determining the position and condition of reinforcement, voids in the concrete or grouting of post-tensioned construction, or variable compaction.

The source will normally be a radioactive isotope enclosed in a portable container which permits a beam of radiation to be emitted. The choice of isotope will depend upon the thickness of concrete involved: Iridium 192 for 25–250 mm thickness, and Cobalt 60 for 125–500 mm thickness, are most commonly used. The beam is directed at the area of the member under investigation and a photograph produced on a standard X-ray film held against the back face. In cases where particularly precise details are required, such as specific identification of reinforcing bars or grouting voids, the image can be intensified by sandwiching the film between very thin lead screens. After development of the film, reinforcement will appear as light areas due to the higher absorption of rays by the high density material, whilst voids will appear as dark areas. If it is required to determine the size and position of reinforcement or defects, photogrammetric techniques can be used in conjunction with stereoscopic radiographs.

Although this technique has become established for examination of steel and voids, it is expensive and requires stringent safety precautions. It is also limited by member thickness; although 600 mm is sometimes quoted as an upper limit, for thicknesses greater than 450 mm the exposure times become unacceptably long.

The National NDT Centre at Harwell (276) has studied use in-situ of cobalt 60 source (1.17 and 1.33 MeV gamma rays) for bridge beams between 300 mm and 800 mm thick, but report that exposure times of several hours may be required. An alternative is to use a 6 MeV Betatron as a switchable X-ray source, providing a higher-power and more effective radiographic facility (277). This has enabled exposure times to be reduced to 15–30 minutes for a section of 600 mm thickness.

8.4.3   *Gamma radiometry*

As in radiography, γ-rays are generated by a suitable radioisotope and directed at the concrete. In this case, however, the intensity of radiation emerging is detected by a Geiger or scintillation counter and measured by electronic equipment. This approach will primarily be used for the measurement of in-situ density of the concrete, although it may also be applied to thickness determination.

As the high-energy radiation passes through concrete some is absorbed, some passes through completely, and a considerable amount is scattered by collisions with electrons in the concrete. This scattering forms the basis of 'backscatter' methods which may be used to examine the properties of material near the surface as an alternative to 'direct' meaurements of the energy passing through the member completely.

The first use of this approach seems to have been in the early 1950s when Smith and Whiffin (278), Hass (279), and others reported applications. More recently, Simpson (280) has described developments of the backscatter technique, whilst Honig (281) has outlined methods adopted in the former Czechoslovakia.

8.4.3.1   Direct methods.   A variety of test arrangements have been adopted, but the basic equipment consists of a suitably housed radioactive source, similar to that used for radiography, together with a detector. The detector will usually consist of a counter housed in a thick lead sheath to exclude signals other than those coming directly from the source. The radiation beam may pass directly through the concrete member as shown in Figure 8.12 or alternatively source and/or detector may be lowered into predrilled holes in the body of the concrete if density variations with depth are required.

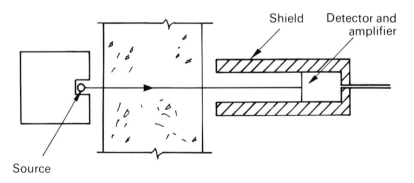

Figure 8.12 Direct radiometry.

Calibrations may be made by cutting cores in the path of radiation after test and using these to provide samples for physical density measurements. As with radiography, the thickness of concrete tested is limited to about 600 mm, which poses a serious restriction to the use of the method. As an alternative to measuring density, this approach may also be adapted to assess member thickness or the location of reinforcing bars.

Computerized tomography analysis methods are being studied (282) to determine if accurate mapping of internal features of concrete elements can be derived from the differential absorption of gamma rays. This is a specialist technique that has yet to be used on full-sized structural elements, but initial laboratory studies are promising.

8.4.3.2 Backscatter methods. It is generally considered that this method tests the density of the outer 100 mm of concrete. The γ-ray source and detector are fixed close together in a suitably screened frame which is placed on the concrete surface. A typical device of this type has source and detector angled at approximately 45° to the surface and at a spacing of approximately 250 mm. In this case the rays propagate through the concrete at an angle to the surface and the intensity of radiation returning to the surface at this fixed distance from the source is measured as illustrated in Figure 8.13.

If density measurements are required at a greater depth, a device consisting of a screened source and detector assembly which may be lowered into a single borehole may be used. Simpson (280) and others have developed calibrated gauges based on the backscatter technique to permit quantitative assessment of the bulk density of concrete, but the results obtained from other forms of equipment can be used for comparative measurements. While this method appears to be simple, difficulties may arise from the non-uniform radiation absorption characteristics of concrete, and inaccuracies may result

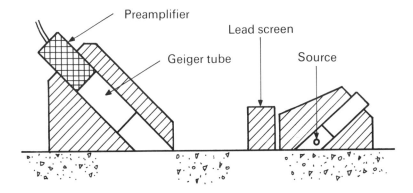

Figure 8.13 Backscatter radiometry.

where there is a density gradient. A number of instruments using the backscatter method are commercially available, with provision also for near-surface direct measurements up to 300 mm deep, with a drilled hole; some instruments also permit neutron moisture measurement (see section 7.2.2).

8.4.3.3 Limitations and applications.   A major problem with γ-radiometry is the expense of the detecting equipment, which requires skilled personnel for its operation, although this has been reduced with the development of commercially available gauges. Scintillation detectors have been favoured in recent years because of the higher maximum count rate which permits the use of a less intense source. The direct method offers the greatest versatility, despite path restrictions, and can be used to detect reinforcement or member thickness in addition to density measurements. Backscatter methods are limited to surface density measurements in most cases. Although valuable in specialized in-situ situations when large numbers of repetitive measurements are required, or as a quality control method for precast construction, radiometry is unlikely to replace conventional gravimetric methods of density determination when the number of specimens is small.

8.5   Holographic and acoustic emission techniques

Attempts have been made in recent years to apply these established techniques to the examination of concrete structures. Both provide means of monitoring the intitiation of cracking under increasing load, and although successful application has so far been reported only for laboratory use, it is possible that they may be further developed for in-situ load test monitoring.

8.5.1   *Holographic techniques*

Holographic techniques provide a method of measuring minute surface displacements by examination of the fringe patterns generated when the surface is illuminated by a light beam and photographed under successive loading conditions, and the results superimposed. Hawkins *et al.* (283) have described the application of these methods to concrete models in the laboratory and have concluded that holographic interferometry, which is capable of measuring out-of-plane displacements of the order of one wavelength of the light used, and speckle holography, which can detect in-plane movements of less than one wavelength, are the most useful. In both cases the elimination of vibrations and rigid body motions is essential. Whilst the latter can be achieved by rigidly fixing the optics to the member under test, the elimination of vibrations poses a serious problem even under laboratory conditions. Although these methods may permit the examination of crack development in the laboratory, the equipment is complex and the practical problems associated with site usage appear to be difficult to overcome. Recent developments in laser technology may however help in this respect.

8.5.2 *Acoustic emission*

8.5.2.1 Theory. As a material is loaded, localized points may be strained beyond their elastic limit, and crushing or microcracking may occur. The kinetic energy released will propagate small amplitude elastic stress waves throughout the specimen. These are known as acoustic emissions, although they are generally not in the audible range, and may be detected as small displacements by transducers positioned on the surface of the material.

An important feature of many materials is the Kaiser effect, which is the irreversible characteristic of acoustic emission resulting from applied stress. This means that if a material has been stressed to some level, no emission will be detected on subsequent loading until the previously applied stress level has been exceeded. This feature has allowed the method to be applied most usefully to materials testing, but unfortunately Nielsen and Griffin (284) have demonstrated that the phenomenon does not always apply to plain concrete. Concrete may recover many aspects of its pre-cracking internal structure within a matter of hours due to continued hydration, and energy will again be released during reloading over a similar stress range.

More recent tests on reinforced concrete beams (285) have shown that the Kaiser effect is observed when unloading periods of up to 2 hours have been investigated. However, it is probable that over longer time intervals the autogenic 'healing' of microcracks in concrete will negate the effect.

8.5.2.2 Equipment. The signal detected by the piezo-electric transducer is amplified, filtered, processed and recorded in some convenient form (Figure 8.14).

Specialist equipment for this purpose is available in the UK as an integrated system in modular form (286), and lightweight portable models may be used in the field. The results are most conveniently considered as a plot of emission count rate against applied load (Figure 8.15).

8.5.2.3 Applications and limitations. It has been reported (50) that as the load level on a concrete specimen increases, the emission rate and signal

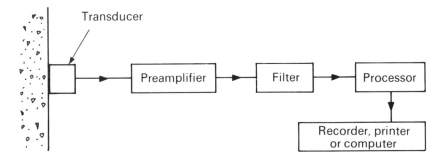

Figure 8.14 Acoustic emission-equipment.

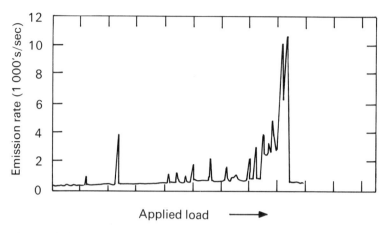

Figure 8.15 Typical acoustic emission plot (based on ref. 50).

level both increase slowly and consistently until failure approaches, and there is then a rapid increase up to failure. Whilst this allows crack initiation and propagation to be monitored during a period of increasing stress, the method cannot be used for either individual or comparative measurement under static load conditions.

Mindess (287) has also shown that mature concrete provides more acoustic emission on cracking than young concrete, but confirms that emissions do not show a significant increase until about 80% to 90% of ultimate stress. The absence of the Kaiser effect for concrete effectively rules out the method for establishing a history of past stress levels. Hawkins *et al.* (283) have, however, described laboratory tests which indicate that it may be possible to detect the degree of bond damage caused by prior loading if the emissions generated by a reinforced specimen under increasing load are filtered to isolate those caused by bond breakdown, since debonding of reinforcement is an irreversible process. Titus *et al.* (288) have also suggested that it may be possible to detect the progress of microcracking due to corrosion activity. Long-term creep tests at constant loading by Rossi *et al.* (289) have shown a clear link between creep deflection, essentially caused by drying shrinkage microcracking, and acoustic emission levels (Figure 8.16). However, in tests on plain concrete beams, together with fibre-reinforced and conventional steel-reinforced beams, Jenkins and Stepotat (290) concluded that acoustic emission gave no early warning of incipient failure.

The application to concrete of acoustic emission methods has not yet been fully developed, and as equipment costs are high they must be regarded as essentially laboratory methods at present. However, there is clearly future potential for use of the method in conjunction with in-situ load testing as a means of monitoring cracking origin and development and bond breakdown, and to provide a warning of impending failure.

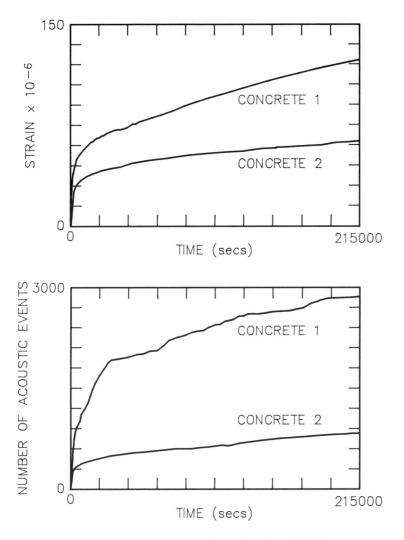

Figure 8.16 Acoustic emission tests on creep specimens (based on ref. 289).

8.6 Photoelastic methods

Photoelastic and similar techniques for tests on concrete members will be
related to load testing or monitoring (Chapter 6). Photoelastic coatings may
be bonded to the concrete surface and will develop the same strains and act
like a continuous full-field strain gauge. Commercially produced sheets of
1–3 mm thickness are available for flat surfaces, and 'contoured' sheeting
may be produced to fit curved surfaces. A portable reflection polariscope
can be used to produce an overall strain pattern, and Redner (291) claims

that measurement of fringes by an experienced operator can yield concrete stress sensitivities of ± 0.15 N/mm^2 in the laboratory or ± 0.35 N/mm^2 under field conditions.

Moiré fringe techniques, in which two sets of equidistant parallel lines are placed close together and their relative movement measured, are also described by Redner (291). A grid is cemented to the structure and compared with a master by reflected light. The technique may be used in similar situations to a photoelastic coating but is restricted to flat surfaces and has a lower sensitivity than photoelastic methods.

The principal use of these methods for the testing of concrete in structures will be to determine patterns of behaviour at stress concentrations, or in complex localized areas under an increasing load or with time. They are best suited to a laboratory environment, although field use is possible.

8.7 Maturity and temperature matched curing

These two techniques may be useful when attempting to monitor performance in terms of in-situ concrete strength development for timing of safe formwork or prop removal, application of loading (including pre-stress) or some similar purpose. Companion test cubes or cylinders stored in air alongside the pour will not experience the same temperature regime as the concrete in the pour, and will usually indicate lower early age strength values. Even if covered by damp hessian or plastic sheeting, test specimens in steel moulds will remain close to air temperature while in-situ concrete temperatures may commonly rise initially by up to 20° or 30°C within the first 12 hours after casting and remain above ambient for several days. The effect of this is particularly critical in cold weather, when the true in-situ strength may be seriously underestimated by companion specimens. Maturity and temperature matched curing approaches both involve measurement of within-pour temperatures to overcome this problem, but location of test points is critical because of the internal temperature variations which will exist.

8.7.1 *Maturity measurements*
Strength development is a function of time and temperature, and for a particular concrete mix and curing conditions this may be related to the maturity, which is defined as the product of time with temperature above a predetermined datum. Maturity is commonly calculated as

$$M(t) = \sum (T_a - T_0)\Delta t$$

where $M(t)$ is maturity in degree-hours or degree-days

Δt is time interval in hours or days

T_a is average concrete temperature during time interval Δt

T_0 is datum temperature.

The datum is the temperature at which concrete is assumed not to gain strength with respect to time and is commonly taken as $-10°C$.

The approach is detailed by ASTM C1074 (292) and multi-channel equipment is commercially available to monitor in-situ temperatures by means of thermocouples or thermistors embedded in the concrete (293). Simple chemical-based devices are also available, as shown in Figure 8.17, which are inserted into the surface of the newly placed concrete (192) but these offer a lower level of precision.

Unfortunately correlation between maturity and strength, as illustrated in Figure 8.18, is specific to a particular concrete mix and curing regime and maturities cannot be used alone because of the risk of variations in the mix composition. Maturity measurements must thus be backed up by some other form of strength testing, but do provide a useful preliminary indicator of strength development.

Carino (294) further discusses in detail the use of concrete maturity to estimate early age strengths. He concludes that the relationship between $M(t)$ and strength for a given concrete mix is sensitive to the early age curing temperature and that a higher temperature will result in a higher early age strength but a lower long-term strength (Figure 8.19). However, there is a unique relative strength vs. maturity relationship (Figure 8.20) which can be used to predict accurately the early age strength as a proportion of the final long-term strength or of the 28 day strength.

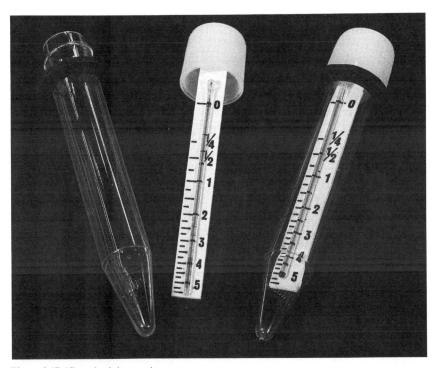

Figure 8.17 'Coma' mini maturity meter.

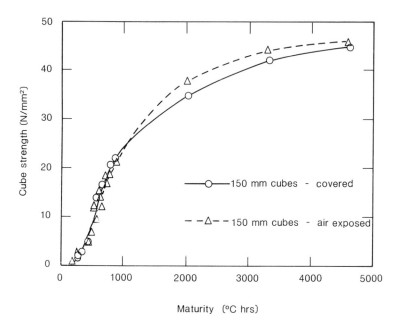

Figure 8.18 Typical strength/maturity results.

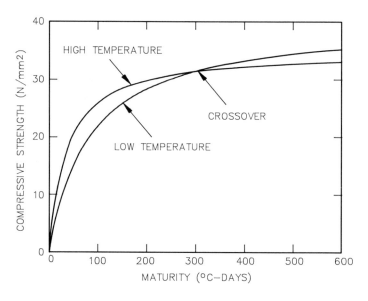

Figure 8.19 The effect of early age curing temperature on the strength–maturity relationship (based on ref. 294).

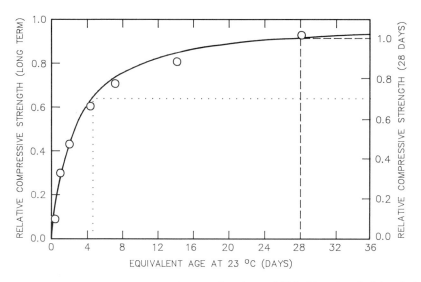

Figure 8.20 Relative strength *vs.* equivalent age (based on ref. 294). (Note that the left scale is based on the final long-term strength and the right scale on the 28 day strength).

The fraction of the 28 day strength, RS_{28} is given by

$$RS_{28} = \frac{S}{S_{28}} = \frac{0.436(t_e - 0.08)}{1 + 0.4(t_e - 0.08)}$$

where S = early age strength
  S_{28} = 28 day strength
  t_e = equivalent age at reference temperature, T_r.

In Europe a reference temperature of 20°C is normally used, whilst in the USA a value of 23°C is usually assumed.

The equivalent age is given by the simplified expression

$$\frac{M(t)}{T_r - T_o}$$

More complex expressions for the equivalent age, giving a greater accuracy, are discussed fully in ref. 294.

In summary, maturity measurements can be used to account for the effects of temperature and time on strength development but cannot be used in isolation to detect batching errors. A potential application is to use the maturity approach in conjunction with supplementary in-situ strength tests to determine whether the as-placed concrete will comply with the contractual strength requirements.

8.7.2   *Temperature matched curing*

The temperature at a pre-selected point within the concrete element may be monitored and used to control the temperature of a water bath in which test specimens (cubes or cylinders) are placed. Their temperature regime, and hence maturity, will thus be identical to that at the selected point in the pour. These specimens may then be tested for strength and may be related to the in-situ properties with due allowance for likely difference in compaction.

The fundamental requirements are detailed by British Standards Draft for Development DD92 (295), to be revised as BS 1881: Part 130 in due course. Features include a tank with sufficient capacity to take at least four test specimens in their moulds and with a water-circulating device and heater capable of producing a temperature rise of at least 10°C/hr. The temperature sensor in the concrete is coupled to control equipment which maintains the water temperature within 1°C of the sensor temperature. A continuous record of both temperatures must be produced to permit conformation of the correct functioning of the equipment and calculation of maturity. Commercially available equipment exists which fulfils these requirements. Harrison (296) and Cannon (297) have both discussed this technique, which is growing in popularity, in greater detail.

8.8   Screed soundness tester

Floor screeds often fail by deformation under a concentrated load, such as a chair leg, and Pye and Warlow (298) have reported a test method to measure soundness of dense floor screeds. The equipment, commonly known as the BRE screed tester, includes a 4 kg weight which is dropped one metre down a vertical rod on to a 'foot' to impact the surface over a 500 mm^2 area. The surface indentation caused by four blows on the same area is measured, using a simple portable dial gauge device. If the indentation is more than 5 mm the screed is unsatisfactory and Pye (299) has indicated classification criteria. These limits will apply to screeds which are at least 14 days old and where the test area is flat and free of all loose dirt and grit. The test must not be used if the screed is laid over a weak or soft insulating material, as a core may be punched out, but is particularly useful in identifying zones of poor compaction beneath an apparently good surface.

8.9   Tests for fire damage

Visual observation of spalling and colour change (9) possibly aided by surface tapping is the principal method of assessment of fire damage. Ultrasonic pulse velocity measurements may perhaps help (300) — see also Chapter 3 — but use on site for this application poses many problems. Surface zone residual concrete strength can be assessed by appropriate partially destructive tests (see Chapter 6), whilst thermoluminescence, as described in section 9.12, may help to determine fire history. Further guidance regarding the assessment of fire damage is given by ACI Special Publication No. 92 (301).

9   Chemical testing and allied techniques

Chemical testing of hardened concrete is mainly limited to the identification of the causes of deterioration, such as sulphates or chlorides, or to specification compliance, involving cement content, aggregate/cement ratio or alkali content determination. Water/cement ratio, and hence strength, is difficult to assess to any worthwhile degree of accuracy, and direct chemical methods are of limited value in this respect. Some chemical tests are expensive, and will often only be used in cases of uncertainty, or in resolving disputes, rather than as a means of quality control of concrete.

Specialist laboratory facilities are required for most forms of chemical testing. Basic procedures for the principal tests are outlined below, and emphasis has been placed on the interpretation and reliability of results. Techniques and procedures are generally complex, and extreme care must be taken both during sampling and testing if accuracies are to be achieved which are of practical value. One of the major problems of basic chemical testing is the lack of a suitable solvent which will dissolve hardened cement without affecting the aggregates, and if possible samples of the aggregates and cement should also be available for testing. Other instrumental techniques, such as differential thermal analysis, requires expensive and complex equipment together with a high degree of skill and experience, but are growing in usage. The range of techniques available to the cement chemist is wide, and many are of such a highly specialized nature that they are outside of the scope of this chapter. Attention has therefore been concentrated on those methods which are most commonly used for in-situ investigations, whilst the more important of the other techniques are indicated together with their most commonly used applications.

ASTM standards are available for commonly used chemical tests, but BS 1881: Part 124 (302) provides more comprehensive guidance and procedural details for many tests. These include cement content, aggregate content and grading, aggregate type, cement type, original water content and bulk density, as well as chloride, sulphate and alkali contents. These procedures apply to calcareous cements, and to natural or inorganic artificial aggregates. Additional background information and details are given by Figg and Bowden (303), and a comprehensive Concrete Society Technical Report offers further detailed guidance (304).

It is particularly important that an engineer requiring chemical analysis of concrete should be aware of the limitations of the methods available, and in particular the effect that some materials' properties may have on the accuracy of analysis. The most likely causes of lack of accuracy are:

(i) Inadequate sampling or testing
(ii) Aggregates contributing to the analysis
(iii) Cements with unusual and unknown composition
(iv) Changes to the concrete from chemical attack or similar cause
(v) Presence of other materials.

It is also essential that an experienced concrete analyst should be employed. He must be given a clear brief of the information required from testing, and all relevant data concerning the constituents and history of the concrete must be made available.

9.1 Sampling and reporting

9.1.1 *Sampling*

The tests undertaken on a sample will relate only to the concrete at that particular location in a structure. It is therefore essential that sufficient samples are taken to represent the body of concrete under examination. Results from particular samples cannot be assumed to provide information on the concrete at points other than those from which the samples were taken. The natural variability of concrete properties as well as the mobility of some chemicals (e.g. chlorides) must thus be considered. At least two, and preferably four, separate samples are recommended for volumes up to $10\,m^3$ of concrete, and at least ten samples should be used for very large volumes under examination followed by more extensive investigation at particular locations if necessary. It is important that, in cases of dispute, agreement is reached between all parties involved about the location and method of sampling, together with the extent of the material which is to be considered to be represented by the sample.

The basic requirements for a sample for chemical analysis are that:

(i) The minimum linear dimension should be at least five times the maximum aggregate size
(ii) The sample should be in a single piece
(iii) The sample should be free from reinforcement and foreign matter unless they are the subject of the test.

The sample should be clearly labelled with the date, precise location and method of sampling with any other relevant details and sealed in a heavy-duty polythene bag, which is also labelled. The size of the sample will vary according to the tests to be performed; some tests may require only a small drilled dust sample whilst others will need a sizeable lump of concrete. Most chemical tests can be carried out on a sample obtained from a core after compressive testing, although if an undamaged specimen is required for original water content determination an untested core will be required. In many cases it will be necessary to determine the cement content of the concrete in addition to the particular component under examination (e.g., water content, sulphate content, etc.). The original sample must therefore be large enough to allow for all the tests to be performed. The quantities of sample subjected to preparation, such as grinding, should also take account of this. It is important that the sampling method does not influence the sample in relation to the property to be measured, for example by washing-out.

9.1.2   *Reporting*
It is desirable that the reports must contain the following information in addition to that required by BS 1881: Part 124 (302):

  (i) Methods used
 (ii) Variation to standard methods, with justification
(iii) Assumptions made, such as aggregate type or properties, with justification
 (iv) Raw analytical data
  (v) Conclusions with a statement of their reliability.

BS 1881: Part 124 further requires that full details of identification marks should be given with a full qualitative description of the sample. Particular reference must be made to factors which are likely to reduce the accuracy of results. Details of other results obtained coincidentally to the principal tests, or any tests done at the analyst's discretion, should also be given.

9.2   Cement content and aggregate/cement ratio

9.2.1   *Theory*
The most common methods for determining the cement content of a hardened mortar or concrete are based on the fact that lime compounds and silicates in Portland cement are generally much more readily decomposed by, and soluble in, dilute hydrochloric acid, than the corresponding compounds in aggregate. The quality of soluble silica or calcium oxide is determined by simple analytical procedures, and if the composition of the cement is known, the cement content of the original volume of the sample can be calculated. Allowance must be made for any material which may be dissolved from the aggregate, and representative samples of aggregate should be analysed by identical procedures to permit corrections to be made. It will not be possible to treat all concretes in an identical manner because of differences in aggregate properties, but the aggregate/cement ratio will be found as well as cement content, whichever method is used.

9.2.2   Procedures

ASTM C85 (305) uses a crushed sample, dehydrated at 550°C for three hours and treated with 1:3 hydrochloric acid. Dissolved silica is determined by standard chemical methods, and the filtrate from this is then tested for calcium oxide content. Calculations depend upon the nature of the aggregate (a sample of which must be available) and figures are provided to assist interpretation.

The techniques recommended by BS 1881: Part 124 (302) are more complex, but follow the same basic procedures with variations in the significance placed upon analytical methods according to the type of aggregate. Aggregates are classified into three broad groups for this purpose:

Type I — Natural aggregates essentially insoluble in dilute HCl
Type S — Natural aggregates mainly soluble in dilute HCl
Type O — Other aggregates.

9.2.2.1   Sample preparation.   The sample is initially broken into lumps not larger than about 50 mm, taking care as far as possible to prevent aggregate fracture. These are dried in an oven at 105°C for 15–24 hours, allowed to cool to room temperature, and divided into sub-samples.

A portion of the dried sample is crushed, ground, and subdivided to provide a powder which passes a 150 μm fine mesh sieve. The recommended procedure is detailed, but care is essential if a representative sample is to be obtained. The operations should also be performed as quickly as possible to minimize exposure to atmospheric carbon dioxide.

The portion is first crushed to pass a 5.0 mm sieve and a subsample of 500–1000 g obtained, which is then crushed to pass a 2.63 mm sieve and quartered to give a sample which is ground to pass a 600 μm sieve. This is also quartered and further ground to pass the 150 μm sieve.

9.2.2.2   Determination of calcium oxide content alone.   A portion of the prepared analytical sample weighing 5 $\pm$ 0.005 g is treated with boiling dilute hydrochloric acid. Triethanolamine, sodium hydroxide and calcein indicator are added to the filtered solution which is then titrated against a standard EDTA solution. The CaO content may be calculated to the nearest 0.1% in this way.

Reliance upon measurements of CaO content alone may be considered acceptable if the calcium oxide content of the aggregate is less than 0.5%, but additional determination of soluble silica content is recommended.

9.2.2.3   Determination of soluble silica, calcium oxide and insoluble residue.   Soluble silica is extracted from a 5 $\pm$ 0.005 g portion of the prepared sample by treatment with hydrochloric acid and the insoluble residue collected by filtering. The filtrate is reduced by evaporation and treated with hydrochloric acid and polyethylene oxide before again being filtered and diluted to provide a stock solution.

The filter paper containing the precipitate produced at this last stage is ignited in a weighed platinum crucible at $1200 \pm 50°C$ until constant mass is achieved, before cooling and weighing. The soluble silica content can be calculated to the nearest 0.1% from the ratio of the weight of the ignited residue to that of the analytical sample.

The calcium oxide content is determined from the stock solution using procedures similar to those in section 9.2.2.2. The insoluble residue is determined from the material retained during the initial filtration process by repeated treatment with hot ammonium chloride solution, hydrochloric acid and hot water followed by ignition in a weighed crucible to $925 \pm 25°C$.

9.2.2.4 Calculation of cement and aggregate content. The cement content should be calculated separately from both the measured calcium oxide and soluble silica contents, unless the calcium oxide content of the aggregate is less then 0.5% (see section 9.2.2.2), or greater than 35% in which case results based on calcium oxide are not recommended. In the latter case, if the soluble silica content of the aggregate is greater than 10% analysis should be undertaken to determine some other constituent such as iron or aluminium compound known to be present in substantially different amounts in the cement and aggregate. This should preferably be present in the larger quantity in the cement.

It is assumed that the combined water of hydration is $0.23 \times$ the percentage cement content, and that 100% oven dried concrete consists of $C\%$ cement $+ R\%$ aggregate $+ 0.23C\%$ combined water of hydration. Thus if

a = calcium oxide or soluble silica content of cement (%)
b = calcium oxide or soluble silica content of aggregate (%)
c = measured calcium oxide or soluble silica content of the analytical sample (%),

then percentage cement content

$$C = \frac{c - b}{a - 1.23b} \times 100 \qquad \% \text{ (to nearest 0.1%)}$$

and percentage aggregate content

$$R = \frac{a - 1.23c}{a - 1.23b} \times 100 \qquad \% \text{ (to nearest 0.1%)}$$

thus the aggregate/cement ratio $= R/C$ to the nearest 0.1.

The cement content by weight is given by

$$\frac{C \times \text{oven dried density of concrete}}{100} \text{kg/m}^3 \text{ to the nearest 0.1 kg/m}^3.$$

If coarse and fine aggregates have differing calcium oxide and soluble silica contents the overall weighted means should be used as the values for b determined on the basis of assessed grading, mix design or visual inspection.

Use of these expressions requires an analysis of both the cement and aggregate to be available. If an analysis for the cement is not available, ordinary and rapid hardening cements complying with the relevant British Standard may be assumed with little loss in accuracy to have a calcium oxide content of 64.5%. An assumed soluble silica content of 20.7% is less reliable. Appendix A of BS 1881: Part 124 (302) provides detailed typical analyses of these and other common cement types. Absence of an aggregate analysis poses greater problems, although in some cases the silica content may justifiably be assumed to be zero (many laboratories are known to use a value of 0.5% silica correction as routine), and for type I aggregates the calcium oxide content may be taken as zero. Otherwise microscopic examination may be necessary (see section 9.11).

If the two estimated cement contents are within $25\,kg/m^3$ or 1% by mass, the value is adopted. If a greater difference exists, and no reason can be found for the discrepancy, both results may be quoted with an indication of the preferred value based on the factors outlined above.

Where pulverized fuel ash or natural pozzolans have been used as reactive components it is difficult to determine the content of the addition, but an approximate estimate may be made of the ordinary Portland cement content provided the added material has a calcium oxide content below about 2%. Mathematical methods involving the use of X-ray fluorescence techniques to determine slag and pfa contents have however recently been reported (306), and other techniques involving chemical analysis of individual grains are believed to be in use. Provided that such additions have not been used, a value of aggregate content may be calculated from the insoluble residue values of the analytical sample and of the aggregate assuming no insoluble residue in the cement. Where the aggregate is type I, the percentage insoluble residue value obtained for the analytical sample may be taken as the percentage cement content.

9.2.3 *Reliability and interpretation of results*

The procedures used above will yield the Portland cement content of a hardened concrete, whilst the content of blended cements can only be determined if the actual analysis of the blend is known. It is however virtually impossible to measure the relative proportions of the blended ingredients. The Concrete Society Technical Report (304) provides details of extensive precision trials involving several test laboratories. Errors may be divided into those caused by sampling, for which $\pm 50\,kg/m^3$ for one sample reducing to $\pm 25\,kg/m^3$ for four samples may be assumed, and those due to testing. Testing accuracy is dependent upon the aggregate type and test procedures, but may be of the order of $\pm 15\,kg/m^3$ for limestone ranging to $\pm 40\,kg/m^3$

for flint aggregates. Combining these effects typically gives an overall accuracy of between $\pm 30\,kg/m^3$ to $\pm 50\,kg/m^3$ based on the mean of four independent samples. A figure of $\pm 45\,kg/m^3$ is often regarded as typical for a gravel concrete with a cement content of the order of $350\,kg/m^3$. This value will be considerably worse for atypical cements, or when blast furnace slag or pulverized fuel ash are present. Whenever assumptions about the contribution of acid-soluble aggregate components must be made because of lack of availablity of a separate sample, the accuracy will decrease. Similarly, assumptions of CaO and SiO_2 proportions in the cement will affect accuracy, but since the range of CaO in a typical UK Portland cement is less than that of SiO_2, assumptions based on CaO should be preferred whenever possible. If the aggregate effect is high, the silica approach may be used, but although this is less aggregate-sensitive it is not so easy or reliable.

The importance of careful sampling and sample preparation cannot be overemphasized. If the aggregates prevent analysis of cement content and aggregate/cement ratio by chemical means, it may be possible to obtain estimates based on micrometric methods outlined in section 9.11, or by instrumental techniques such as X-ray fluorescence spectrometry.

Whatever method is used to determine cement content, it is essential to recognize that no statistical statement about a particular batch of concrete can be made from a test on a single sample. Where several separate samples are used, an estimate of an average value may be possible, and a minimum of four individual samples should be used to obtain a reliable estimate. Selection of the location for these should preferably take account of variations caused by position in mixer discharge, bearing in mind that for a truck mixer the first portion will generally be richer than average whilst the later portions may be leaner. Differences of up to $100\,kg/m^3$ have been reported (304) between the top and bottom of walls and columns.

9.3   Original water content

9.3.1   *Theory*

The quantity of water present in the original concrete mix can be assessed by determining the volume of the capillary pores which would be filled with water at the time of setting, and measuring the combined water present as cement hydrates. The total original water content will be given by the sum of the pore water and combined water. This approach was originally developed by Brown (307) and forms the basis of the BS 1881: Part 124 method (302). It requires a single sample of concrete which has not been damaged either physically or chemically. Usually the water/cement ratio is of greatest interest, so that a cement content determination will also be necessary. Alternative methods of assessing water/cement ratios include thin section microscopy and reflected light fluorescence microscopy (308).

9.3.2 *Procedure*

An undamaged sample is normally obtained as a saw-cut slice approximately 20 mm thick and with a single face area of not less than 10 000 mm^2 (e.g. 100 mm square). The sample should be taken from sufficient depth within the concrete to avoid laitance and other surface effects, and may conveniently be obtained as a 'vertical' slice on a core. Care must be taken to minimize material loss from the cut faces, and carbonation is prevented by storage in an airtight container.

The sample is oven dried at 105°C for at least 16 hours, cooled in a desiccator, weighed, and immersed in a liquid of known density (commonly trichlorethane), in a vacuum desiccator. The pressure is reduced, causing the air from the capillaries to be evolved. The vacuum is then released and the sample kept immersed in the liquid for a further five minutes, then weighed in a sealed polythene bag to prevent loss by evaporation. The weight of liquid filling the pores can thus be calculated, and the % capillary water derived:

$$\% \text{ capillary water} = \frac{\text{wt. of liquid absorbed}}{\text{liquid density} \times \text{wt. of dry sample}} \times 100.$$

After the capillary porosity measurements are complete the slice is heated at 105°C until constant weight is obtained, and crushed to pass a 150 μm sieve. Approximately 1 g of this sample is ignited at 1000°C in a stream of nitrogen or dried air, and the evolved water is weighed after it has been absorbed by 'dried' magnesium perchlorate. A further portion of this ground sample is used for the cement content measurement, using the most appropriate method of those given in section 9.2, according to aggregate properties.

Unfortunately the aggregate will frequently also have a porosity and combined water content that must be allowed for. These can be assessed by applying identical procedures to those above to a sample of coarse aggregate.

It must be assumed that the coarse aggregate values are typical for all aggregates in the concrete slice, and hence

$$\text{corrected capillary water} = Q - \frac{qR}{100}\%$$

$$\text{and corrected combined water} = X - \frac{YR}{100}\%$$

where Q = determined capillary porosity of slice, %
q = determined capillary porosity of aggregate, %
R = aggregate content, %
X = combined water of concrete, %
Y = combined water of aggregate, %
C = cement content, %.

If it is assumed that no water of hydration has been replaced by carbon dioxide, the original free water/cement ratio is given by the sum of the above expressions divided by the cement content $C\%$. The original total water/cement ratio can similarly be derived using the uncorrected percentage capillary porosity Q with the corrected percentage of combined water.

If aggregate control samples are not available it must be assumed that the combined water of hydration is typically 0.23 times the percentage cement content. This value is often adopted as a matter of routine because of inaccuracies which are likely in combined water measurements. The capillary porosity of the aggregate may also be taken to be equal to the water absorption value of the aggregate, if this is known, thus permitting an estimate of original free water/cement ratio. If the aggregate absorption value is not known, then only the original total water/cement ratio can be quoted and is given by $(Q/C) + 0.23$.

9.3.3   *Reliability and interpretation of results*

Since this determination requires a sound specimen of concrete, strength-tested cores or cubes cannot be used. It is suggested that a precision better than ± 0.1 for the original water/cement ratio is unlikely to be achieved even under ideal conditions (304). A major source of difficulty lies in the corrections for aggregate porosity and combined water which may be overestimated by the procedures used. This will lead to an underestimate of the true original water content although it is also possible that continued hydration in older concretes may lead to low apparent original water/cement ratios. The original cement content must also be measured with reasonable accuracy, and the problems associated with this are discussed in section 9.2.

The method is not suitable for semi-dry or poorly compacted concrete, although in the case of air entrainment, Neville (309) suggests that since the voids are discontinuous they will remain air-filled under vacuum and will absorb no water although this view is regarded by others as doubtful. If this is the case, entrained air will not affect the results, which are influenced only by capillary voids. Carbonated concrete must also be avoided. Where the aggregates are very porous or contain an appreciable amount of combined water the corrections required will be so large that the results may be of little value. This shortcoming is likely to limit the value of the method for artificial aggregates.

9.4   Cement type and cement replacements

9.4.1   *Theory*

The basic type of cement used in a concrete may be established by separation and chemical analysis of the matrix, for comparison with established analyses of particular cement types. Alternatively, microscopic examination can be used to detect unhydrated cement particles which can be compared with known specimens. The complexity of the analysis varies according to cement type, admixture, replacements and aggregates, and BS 1881: Part 124 (302) strongly recommends that chemical analysis should be supplemented by microscopic examination. It is not possible to distinguish between ordinary and rapid hardening Portland cements, although cement replacements can usually be detected. Complex techniques such as differential thermal analysis may also be used to establish cement type on a comparative basis. A simple site test has been developed for high alumina cement identification (see section 9.4.2.4).

9.4.2 Procedures

9.4.2.1 Analysis of matrix. A very fine sample is obtained by sieving material from a broken piece of concrete through a 90 μm mesh. It is essential that this contains no more than minute traces of aggregate which may contribute to the analysis. The very fine sample can then be analysed for basic cement components such as SiO_2, CaO, Al_2O_3, Fe_2O_3, MgO and SO_3 and the resulting composition compared with analyses of known cements.

9.4.2.2 Microscopic method. A small solid piece of concrete, about 20 mm cube, should be selected, preferably not containing any large aggregate pieces nor having a finished surface. This is dried for 12 h at 105°C and embedded in epoxy mortar before saw cutting and grinding with carborundum powder. The surface is finally polished with diamond powder and examined by reflected light microscopy. Coarse unhydrated cement particles will be visible, and these may be etched to show the phases characteristic of cement type. BS 1881: Part 124 (302) offers guidance concerning the most appropriate etching methods, and comparative reference specimens of known cement type may be valuable for identification. Power and Hammersley (310), Petersen and Poulsen (192) and French (311) have also described approaches using thin microscopic sections.

9.4.2.3 Cement replacements. The most commonly used cement replacements are blast furnace slag and pulverised fuel ash (pfa). Blast furnace slags contain considerably higher levels of manganese and sulphides than are to be found in normal cements, and this can be used for their identification by chemical analysis. If the slag is from a single source of known composition it may be possible to obtain quantitative data, provided there is no slag in the aggregate. A characteristic green or greenish black coloration of the interior of the concrete may aid identification, and microscopic methods may also be used. Pfa is most easily detected by microscopic examination of the acid-insoluble residue of a sample of separated matrix. Particular characteristic spherical particle shapes may be recognized, although quantitative assessment of the pfa content is not possible. Microscopic examination of thin sections as described by Power and Hammersley (310) and French (311) may also be used.

9.4.2.4 Identification of high alumina cement (simple method). A simple rapid chemical test for the identification of high alumina cement has been developed by Roberts and Jaffrey (312). Whilst this type of cement can be identified by microscopic or complex methods such as X-ray spectrometry, the need for a simple test that could be used on site arose following problems with this material in the UK. This method is based on the assumption that an appreciable quantity of aluminium will be present in HAC concrete dissolved in dilute sodium hydroxide, whereas little will be found in a similar solution of Portland or other types of cement.

A powdered sample is obtained by drilling by masonry drill, or by crushing small pieces of concrete with a pestle and mortar and grinding after removal of aggregate particles. Approximately 1 g of this powdered sample is placed in a 25 ml test tube with about 10 ml of 0.1 N cold sodium hydroxide solution, sealed and then shaken by hand for 2–3 minutes. The solution is then filtered through a medium-grade paper and the filtrate acidified with five drops of dilute HCl. Ten drops of 'Oxine' and 1 ml ammonium acetate are then added, and if aluminium is present in appreciable quantities, as in HAC, turbidity and the formation of a yellow precipitate will occur. If the solution remains clear or only slightly cloudy, HAC is not likely to be present. The presence of gypsum, which may result from contamination of the sample by plaster, can interfere with the test and produce misleading results. Care must therefore be taken to avoid such contamination during sampling.

9.4.3 Reliability and interpretation of results

Both chemical and microscopic methods of cement type determination rely upon a carefully prepared sample and a great deal of specialist skill and experience. Whilst under favourable conditions it should be possible to identify cement type, neither method is accurate enough to tell whether that cement complied with a particular specification. Cement replacements can usually be identified, but the level can only be determined for blast furnace slag, and then ony if conditions concerning the source and aggregate properties are favourable. The simple HAC determination method has been developed for site use and has proved to be a reliable indicator of the presence of this type of cement provided that sampling is carefully executed.

9.5 Aggregate type and grading

9.5.1 *Aggregate type*

This is best determined in a cut slice or core which will allow the cut sections of aggregate to be examined visually, physically or chemically to determine its type. ASTM 856 (313) provides guidance on petrographic examination. The description and group classification of aggregates is dealt with in BS 812: Part 1 (314) but petrographical examination by a qualified geologist is required. If a more detailed analysis relating to mineralogy, texture and microstructure is needed microscopic methods described by Power and Hammersley (310) and French (311) may be useful. These can yield information on hardness, porosity, permeability, specific gravity and thermal properties of the aggregate as well as the presence of potentially deleterious substances. This is also the only reliable method of distinguishing between frost attack and alkali/aggregate reactions as the cause of deterioration of hardened concrete.

In many instances a detailed aggregate identification will be unnecessary, and a broad chemical classification based on reaction with acid will be adequate. This will apply particularly to the selection of the analytical method to apply to the concrete for cement content determination.

9.5.2 *Aggregate grading*

This can only be reliably achieved for aggregates which are essentially insoluble in dilute hydrochloric acid. An initial sample of at least 4 kg will be necessary, and a sub-sample of approximately one quarter of this will be tested. This is initially broken down into coarse and fine fractions using a 5 mm sieve without fracturing the aggregate, assisted as necessary by heating. Coarse aggregate particles are cleaned by chipping and both fractions treated with hydrochloric acid to dissolve the paste prior to sieve analysis. Care must be taken to remove chipped portions of coarse aggregate from the fines, and gradings obtained are not of sufficient accuracy to assess compliance with detailed aggregate specifications.

9.6 Sulphate determination

The sulphate content is obtained by chemical analysis of a weighed, ground sample of concrete which is expected to contain about 1 g of cement. The use of a concentrated sample of fine materials, obtained as described in section 9.2.2.2 may be worth while. This is treated with HCl, ammonia solution and barium chloride solution to obtain a precipitate which is ignited at 800°C–900°C and weighed. If the sulphate content of the original sample exceeds about 3% of the cement content for Portland cement concretes, chemical attack may be indicated. A cement content determination must therefore be performed in conjunction with sulphate tests, and an accuracy better than $\pm 0.3\%$ SO_3/cement is unlikely. Specialized instrumental techniques (see section 9.13) are growing in popularity and may improve on this accuracy.

Cases of suspected sulphate attack may be identified by measuring the sulphate distribution between sulphoaluminate and gypsum. Since sulpho-aluminate is stable only in alkaline conditions and is converted to gypsum by acid, attack by groundwater may be detected and distinguished from the products of suphur-oxidizing bacteria.

Sulphate attack may also be identified by petrographic methods (see section 9.11) involving the microscopic study of thin sections to reveal the presence of crystalline calcium sulphoaluminate. This is commonly known as ettringite and may be found along cracks when reactions have occurred between sulphate-bearing solutions and the cement paste. When high alumina cement is used, complex methods such as differential thermal analysis may be required (see section 9.13).

9.7 Chloride determination

The need to assess the chloride content of hardened concrete is most likely to arise in relation to the corrosion risk to embedded reinforcement or ties. Small quantities of chloride (up to approx. 0.01% chloride ion by weight of concrete) will normally be present in concrete, but substantially more can result if calcium chloride (as an admixture) or sea-dredged aggregates are used, and may present a potential hazard. Chlorides may also be absorbed from the surface, as in the presence of sea water or de-icing salts, or enter through cracks in the concrete. In some parts of the world, such as the Middle East, contaminated sand may also present a major problem.

9.7.1 *Procedures*
In addition to the 'Volhard' laboratory method described in BS 1881: Part 124 (302), X-ray fluorescence spectrometry and other sophisticated techniques may be used to determine chloride content. Simplified methods are also available which may be suitable for site use, athough the accuracy of these will be lower than is possible from a detailed analysis. The Building Research Establishment has outlined procedures in information sheets (315, 316) for two common methods, and other available methods have been compared by Figg (317).

Whichever analysis technique is to be used, a powdered sample of concrete is required with a total weight of at least 25 g. This can be obtained by the methods described previously (section 9.2.2), usually from sliced cores or else by drilling with a masonry bit in a hand-held slow-speed rotary percussion drill (Figure 9.1). Care must be taken to avoid steel, which may be located by Covermeter. Views differ concerning sampling techniques using drilling

Figure 9.1 Field collection of drill dust sample.

(317) although it is generally accepted that a drill diameter at least equal to the size of the coarse aggregate is necessary to obtain a representative sample, and that a similar depth for sample increment is also required. 20 mm diameter is thus often used but some engineers prefer to combine drillings from several adjacent smaller diameter holes. If surface layer concentrations are required shallower depth samples or increments should be used, but if results are intended to relate to the interior of a body of concrete the drillings from a 5 mm surface zone should be discarded. Externally exposed surfaces may have been leached by rainwater and give unrepresentative results, so if shallow drilling is used it should preferably be done from the inside of a building but avoiding contamination from gypsum plaster or other materials. Surface zones exposed to sea water penetration may similarly give unrepresentative results in relation to the body of concrete.

Drillings can be collected in a clean container pressed against the concrete surface below the drill hole, with care taken to avoid the loss of fine material. It is most important that all the drill dust is carefully collected and is not allowed to blow away under windy weather conditions. Where chlorides are present in the sample, the fine dust will often contain a significantly higher level of chloride than the coarse dust. A variety of funnels and collection tube devices has been developed (317) to assist sample collection. Vacuum drilling techniques are also available, which prevent loss of fine particles from the sample.

9.7.1.1 Laboratory analysis (Volhard method). This method, which is detailed in BS 1881: Part 124 (302), is relatively simple and reliable but does require specialized laboratory facilities and experience. A weighed portion of a ground or powdered sample which is expected to contain about 2 g cement is treated with hot nitric acid and then diluted, filtered and cooled. Silver nitrate and nonylalcohol are added with ferric alum indicator and titrated with ammonium thiocyanate. It is usual to express the results as % chloride ion by weight of concrete or cement.

9.7.1.2 Instrumental methods. X-ray fluorescence spectrometry (section 9.13) may be used to determine both chloride and cement content of compressed samples. The wavelength and intensity of fluorescent radiation generated by bombardment with high-energy X-rays is compared with the characteristics of standard samples of known composition. The method requires specialized sample preparation and test equipment, but analysis is quick, and chloride and cement content are obtained from the same sample, which is available for a repeat analysis if that should be necessary. More rapid methods such as ion chromatography and use of ion-selective electrodes are growing in popularity and a rapid site version of the latter approach is available (13).

9.7.1.3 'Hach' simplified method. This makes use of a commercially available kit (318) for a drop count titration with silver nitrate solution in the presence of potassium chromate indicator. The kit includes a bottle of silver nitrate soution with a drop dispenser, capsules of indicator, a plastic measuring tube and a glass bottle for the titration. A powdered 5 g sample is weighed and dissolved in 50 ml of 1 N nitric acid and 5 g sodium bicarbonate is dissolved in this solution which is decanted and filtered. A 5.75 ml sample of filtrate is measured and placed in the titration bottle with the contents of an indicator capsule, and the silver nitrate solution is added drop by drop until there is a colour change from bright yellow to a faint reddish brown (Figure 9.2) (see reference 315 for a detailed procedure).

For a test on one 5.75 ml measure of filtrate, the percentage chloride ion by weight of concrete is given directly by the number of drops × 0.03, whilst if less than four drops are required two additional measures of filtrate will be added and the total number of drops required by the three measures recorded. In this case the required result is given by total number of drops × 0.01. This method is straightforward and quick, and although a working location is required, should be suitable for site use by staff without specialist experience.

9.7.1.4 'Quantab' simplified method. This method uses a commercially available 'Quantab' test strip (319) to measure the chloride concentration of a solution. The solution, containing 5 g powdered concrete, is obtained as

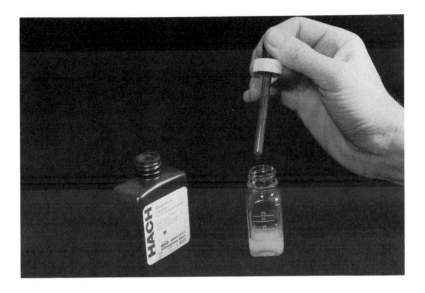

Figure 9.2 Titration using the Hach method.

described in section 9.7.1.3. The strip is plastic, approximately 75 mm long and 15 mm wide, with a vertical capillary column impregnated with silver dichromate (see Figure 9.3). At the top of the column is a horizontal air vent containing a yellow moisture-sensitive indicator which changes to blue when the capillary is full. The lower end of the test strip is placed in the chloride solution until the capillary is full, and the reddish brown silver dichromate in the capillary tube reacts with the chloride to form white silver chloride. The tip of this colour change is related to a vertical scale, and the reading converted to mg chloride ion/litre by reference to calibration tables. The procedure is discussed in detail in reference 316 and the use of 'low range' test strips (Type 1175) for normal purposes is recommended. Caution should be taken with the 'high range' strips which may overestimate chloride concentrations in some situations. Although facilities for sample preparations are required, the method should be suitable for site use by staff without specialist experience, and will be of sufficient accuracy to indicate the presence and level of significant chloride contents for most practical purposes. The total time required for a test is approximately 30 minutes, and in the authors' experience this has been found to be the most reliable of the simplified methods.

9.7.2 *Reliability and interpretation of results*

Chloride distribution may vary considerably within a member due to migration and other effects and an adequate number of samples must be obtained. The Building Research Establishment (315) recommends single

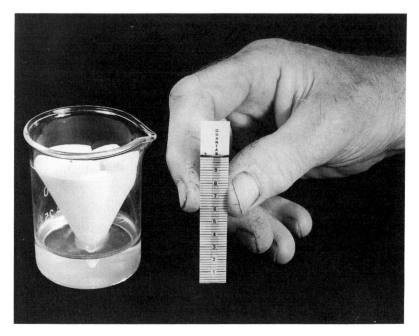

Figure 9.3 'Quantab' chloride test.

samples from at least 10% of a group of building components under investigation, for the identification of the presence of significant quantities of chloride within members of the group. If an individual member is under examination, a number of samples should be taken according to size.

In most instances, the main requirement will be to establish the presence of levels of chloride higher than would be normally expected and the simplified methods will be adequate. The site use of 'Quantab' test strips is quick and can determine concentrations within the range 0.03% to 1.2% chloride ion by weight of concrete, or establish if even less is present. A similar accuracy may be possible with the 'Hach' method. Results expressed in the form of % chloride ion by weight of concrete will be adequate for the simplified methods, and should be used if big differences are expected between the mixes specified and obtained. In doubtful or borderline cases, or where the precise level is required, proper laboratory determinations will however be necessary. In these cases the cement content of the concrete must also be determined as in section 9.2, so that the result can be expressed as percentage chloride ion by weight of cement. Figure 9.4 shows a comparison of results using the Volhard, Quantab and Hach methods for a series of laboratory specimens when a known percentage of sodium chloride by weight of cement was added to the original concrete mix. It is very important that the basis of presentation of results is clearly indicated to avoid confusion at the interpretation stage.

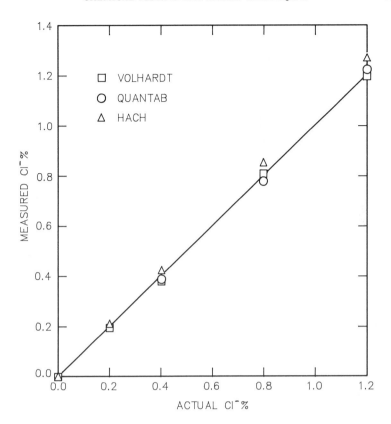

Figure 9.4 Comparison of chloride analysis results.

If the cement content is unknown, a typical value of 14% may be assumed, and hence the percentage chloride content by weight of concrete is multiplied by a factor of seven to estimate the relationship to cement (316). Even if the cement content ranges between 11%–20%, the accuracy should be within ±0.2% using this assumption for chloride contents which are less than 0.10% by weight of concrete. A precise chloride determination will normally be necessary as the referee method in disputes, but considerable savings of cost and effort may be made by first using the simplified methods which do not require expensive equipment or experience. Whichever method of analysis is used, when obtaining samples by drilling, large aggregate particles should be avoided, since these may lead to an underestimate of the true percentage of chloride in the concrete. It is difficult to improve on an overall accuracy of ±0.05% chloride ion/cement at a 95% confidence level, whatever method is used, due largely to sampling errors.

BS 8110 (320) suggests that the maximum acceptable percentage of chloride ion by weight of cement is 0.1% for prestressed concrete, or upper 95% confidence limit of 0.4% for reinforced concrete. Deterioration caused by chlorides is discussed in detail by Kay *et al.* (321) who also suggest that a value of 0.4% chloride ion by weight of cement may be sufficient to promote corrosion of reinforcement, although it is now generally accepted that only 0.25% chloride ion by weight of cement ingressing into hardened concrete is required to depassify the steel reinforcemeent. Whilst these values are generalized, they may assist the interpretation of results on chemical analysis.

9.8   Alkali reactivity tests

With the increased concern in the UK about the risks of alkali-silica reaction (ASR) causing damage to concrete (see section 7.4), tests for alkali content and alkali induced expansion have been developed.

9.8.1   *Alkali content*

An analysis method for hardened concrete has been incorporated into BS 1881: Part 124 (302) which involves the use of a calibrated flame photometer to assess the sodium oxide and potassium oxide contents of a treated powdered sample. A number of uncertainties exist with this method, including the possible contribution of sodium and potassium compounds contained in aggregates but which are not readily available for reaction in concretes. It is common practice to remove as much as possible of the coarse aggregate and to analyse a fines-rich fraction to minimize this problem. Migration or leaching of alkalis within test specimens during sampling or storage is a further cause of uncertainty. Precision and accuracy data for this approach are not currently available.

9.8.2   *Alkali immersion test*

This is intended to give a rapid indication of the presence in a concrete sample of some potentially alkali-silica reactive particles, but does not prove the past or future occurrence of damaging ASR in the concrete. The test procedure is described in detail by Palmer *et al.* (23). A diamond-cut surface is lapped flat, and the sample fully immersed in a 1 N concentrated alkali solution taking appropriate safety precautions. Precautions are also necessary to prevent evaporation during the test period, which is usually up to 28 days at 20°C, although this may be accelerated by higher temperatures.

Small gel growths on the prepared surface may be monitored visually, and any reactive particles identified. Comparative photography before and after testing is essential to confirm and record results.

9.9   Admixtures

Admixtures are frequently organic and are not easy to identify without the aid of sophisticated equipment such as infrared absorption spectrophotometers or high pressure liquid chromatographs. Experience in the application of these to concrete technology is limited, and assessment of the dosage used (which is frequently the prime object of the test) is dependent upon a precise knowledge of the admixture.

9.10   Carbonation

Carbonation of concrete by attack from atmospheric carbon dioxide will result in a reduction in alkalinity of the concrete, and increase the risk of reinforcement corrosion. This will normally be restricted to a surface layer of only a few millimetres thickness, in good quality concrete (322) but can be much deeper in poor quality concrete, with results as high as 30 mm being not uncommon. The extent of carbonation can be easily assessed by treating with phenolphthalein indicator the freshly exposed surfaces of a piece of concrete which has been broken from a member to give surfaces roughly perpendicular to the external face. Alternatively, incrementally drilled powdered samples may be sprayed or allowed to fall on indicator-impregnated filter paper. Drilled cores may also be split and sprayed with indicator to show the carbonation front. A purple-red coloration will be obtained where the highly alkaline concrete has been unaffected by carbonation, but no coloration will appear in carbonated zones.

The colour change of phenolphthalein corresponds to a pH of about 9 whilst reinforcement corrosion may possibly commence at a pore solution pH of about 11. Thus the carbonation front indicated must be recorded as approximate in relation to steel corrosion. The width of carbonation front between these values is of the order of 2–3 mm (323), but an overall accuracy of ± 5 mm is sometimes quoted for a single test using this approach. Accuracy can be improved by taking the mean of several readings, for example five readings can give an accuracy of ± 2 mm at the 95% confidence level. Multicoloured 'Rainbow' indicators are also available (13) which may provide slightly more detailed information.

A number of practical difficulties may sometimes arise as discussed by Theophilus (323). These include:

(i) Freshly broken concrete giving an initially clearly defined colour change boundary may all become coloured within about 1 minute, making the boundary indistinguishable.

(ii) White concrete may immediately register entirely uncarbonated, despite microscopic studies showing a significant carbonated depth. (This is less common.)

(iii) Freshly broken concrete initially registers entirely carbonated, but on subsequent standing the pink coloration appears.

Doubtful results may often be clarified by spraying the surface with deionized water immediately prior to spraying with phenolphthalein, and excessive quantities of indicator should be avoided. Only the finest mist spray is required.

Direct chemical tests may be used to determine the carbonation front with greater precision and may be justified in critical situations. These include the measurement of evolved carbon dioxide from slices of cores, about 5 mm thick (not for calcareous aggregates), using a range of specialized techniques such as thermogravimetry.

Microscopy is probably the most precise method of measuring carbonation, including localized effects of surface cracks (192, 324). Thin sections are viewed in cross-polarized light to reveal calcium carbonate crystals, but calcareous fines may hinder identification of carbonation.

The progression of carbonation with time is often estimated by the expression

$$D = K\sqrt{t}$$

where D is the depth in mm at age t years and K is a constant for the concrete and conditions prevailing (324). The use of this expression to predict the time for the carbonation front to reach a specified depth will magnify the effects of the accuracy range to which the measured carbonation value has been obtained. Watkins and Pitt Jones (325) have suggested that this above expression is a simplification and that a more reliable relationship is

$$D = Kt^x$$

where x lies between 0.5 and 1.0. Good agreement has been shown for extensive site data from Hong Kong with K and x varying according to strength level. Carbonation rates are usually higher in dry or sheltered external concrete than in that exposed to rain; thus differing faces of a structure or element are likely to show different carbonation depths. These factors have been discussed more fully by Somerville (2), whilst assessment of resulting corrosion risk is summarized by Sims (324).

9.11 Microscopic methods

These fall into two basic categories: namely, methods involving examination of a prepared concrete surface by reflected light and those requiring a 'thin section' to be obtained. These, and other petrographic methods, are considered in detail by the Concrete Society (304) and by French (311).

9.11.1   *Surface examination by reflected light*

Major internal crack patterns caused by alkali-silica reactions can be examined by viewing under ultra-violet light a polished cut concrete surface which has been sprayed or impregnated with a fluorescent dye. A portable field inspection kit is also available for use on a treated in-situ surface (326), but care is needed to allow for weak fluorescence which may be caused by carbonated concrete (327).

Specialized laboratory techniques can also be used to determine cement, aggregate, and air content of samples taken from in-situ concrete.

9.11.1.1   Theory.   A varnished sawn face of a dried concrete specimen can be examined by stereomicroscope to give the volumetric proportions of a hardened concrete. Polivka (328) has described a 'point count' method based on the principle that the frequency with which each constituent occurs at equally spaced points along a random line on the surface will reflect the relative volumes of the constituents in the solid. This is because the relative volumes of the constituents of a heterogeneous solid are directly proportional to the relative areas on a plane section, and also to the intercepts of these areas along a random line on the section. An alternative approach is the linear traverse technique, in which the intercepts of the constituents are measured along a series of closely spaced regular transverse lines. In either case the aggregate and voids can be identified, and the remainder is assumed to be hydrated cement. The total volume of this can be computed and converted to a volume of unhydrated cement if the specific gravity of the dry cement and non-evaporable water content of the hydrated cement are known.

9.11.1.2   Procedure.   The cut surface must be carefully prepared so that the constituents are readily distinguishable. This is best achieved by grinding, polishing, and impregnation with a suitable dye before varnishing. The prepared sample will normally be examined by a stereomicroscope with a travelling specimen stage, which may be manually or motor driven. Counting will be manually controlled but may conveniently be linked to a microcomputer which is coupled to the moving stage to record its location automatically. A magnification of $50 \times$ is generally used.

ASTM C457 (228) concerns the measurement of entrained air using either the linear traverse or modified point count techniques. In the linear traverse method, records are kept of the total number of sections of air voids, the total distance across air voids, and the total distance across the remainder. The modified point count method involves the recording of the frequency of each component coinciding with a regular system of points, coupled with the frequency of intersection of each component by regularly spaced lines.

Both methods permit calculation of the number, size and spacing of the air voids to be determined. The size is usually expressed in terms of the specific surface, whilst the commony used 'spacing factor' represents the maximum distance from any point in the paste to the periphery of an air void. Markestad (329) has recently described the application of this approach to measurement of air voids in hardened concrete from offshore platforms, and also gives details of the use of an automatic image-analysing microscope for this work. This has the advantage of being much quicker than the tedious standard microscopic techniques. The optical image of a point on the ground surface is projected onto the photosensitive surface of a TV camera tube and converted to an electrical signal. The position and strength of each signal can be classified and stored for processing, and a data printout gives the area or volume percentage, and specific surface, of air voids together with details of scanning of a particular area. Special surface preparation is necessary to fill the pores with white gypsum whilst the remainder of the surface is blackened.

9.11.1.3   Reliability and applications.   It is claimed that this method can measure the cement content to $\pm 10\%$, and the total aggregate content and coarse/fine aggregate ratio can be assessed similarly. However, the total water content cannot be assessed, since water and air voids cannot be distinguished, except for entrained air voids which will be identifiable by their spherical nature and uniformity of size. In situations where cement content cannot be determined chemically this approach may be valuable. Details of precision experiments are given by the Concrete Society (304).

This method has also become accepted for measurement of air entrainment. The choice between point count or linear traverse techniques will depend on circumstances, but Markestad (329) has indicated problems of component identification leading to uncertainties with the point count approach. Whichever method is used, the procedures are tedious and require specialized equipment and skill both in sample preparation and measurement. Modern electronic aids have eased the burden of data collation, and automatic image analysing microscopes will probably be more widely used.

9.11.2   Thin-section methods

Applications of microscopic examination of thin sections have been outlined in earlier sections of this chapter, and include identification of mix components (304), carbonation, and causes of deterioration. Considerable growth in usage of this approach has been experienced in connection with alkali-silica reaction where the method is invaluable in confirming that reaction has occurred, examining the size and extent of cracks, and identifying reactive aggregate particles (192) as illustrated in Figure 9.5.

Sample preparation involves cutting a slice of concrete from a core by diamond saw (preceded if necessary by vacuum resin impregnation), drying and impregnation by low viscosity epoxy resin. This will then be cut and ground using standard petrographic procedures to a $30\,\mu$m thickness using

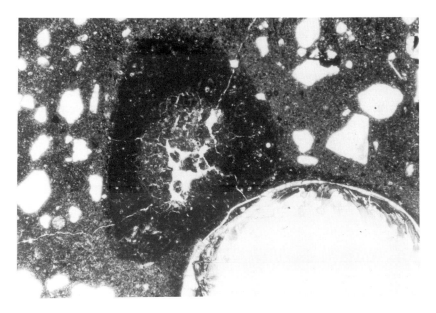

Figure 9.5 Photomicrograph of alkali-silica reaction (photograph by courtesy of Professor E. Poulsen). Note: An opaline chert particle is seriously affected by ASR, with cracks radiating into the surrounding paste. The air void in the lower right hand corner is also partially filled with gel.

oil lubrication to avoid the dissolution of water-soluble materials. Detailed procedures have been described by Poulsen (330), and are summarized by Palmer (23).

Samples will typically be examined with a petrographic microscope under ordinary and polarized light. Micrographs may be produced for record purposes and to illustrate interpretation, which is highy specialized.

9.12 Thermoluminescence testing

It has been proposed by Placido (331) that the thermoluminescence of sand extracted from concrete can form the basis of a test for fire-damaged concrete which is a measure of the actual thermal exposure experienced by the concrete.

9.12.1 Theory

Thermoluminescence is the visible light emission which occurs on heating of certain minerals, including quartz and feldspars, and it is known that the curve of light output vs. temperature for a given sample depends upon its thermal and radiation history. This forms the basis of established techniques of mineral identification, radiation dosimetry and pottery dating. In naturally occurring quartz sand, this light emission occurs within the temperature range of 300°C–500°C, but if samples are reheated there is no emission of

light up to the temperature of the preceding heating. The subsequent pattern
of light emission has also been shown to depend on the period of exposure
to the particular temperature (Figure 9.6).

9.12.2 Equipment and procedure

The equipment required is somewhat complex, including a servo-controlled
glow oven to heat the sample, of only a few mg, in an oxygen-free nitrogen
atmosphere. Radiation is detected by a photomultiplier and fed into a
photon-counter which drives the Y axis of an X-Y recorder whilst the X axis
is driven by a thermocouple monitoring the heating plate temperature. The
small sample required should be drilled by slow drilling with a small
battery-powered masonry drill in order to reduce drill bit temperatures, and
the sample should be washed in concentrated acid solutions to remove
minerals which may give spurious outputs. The remaining quartz sand may
then be tested, although if there is none present it may be possible to utilize
quartz from coarse aggregate, or to modify procedures to use other minerals
which may be present in the aggregates.

9.12.3 Reliability, limitations and applications

The practical value of this test for concrete assessment hinges upon the fact
that the emission temperature range of 300°C–500°C is critical in the

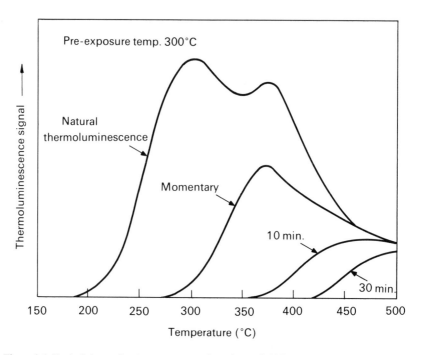

Figure 9.6 Typical thermoluminescence trace (based on ref. 331).

deterioration experienced by concrete subjected to fire, and it is also believed that loss of strength is influenced by thermal exposure as well as maximum temperature. Whilst little data relating thermoluminescence results to concrete strength are available, Chew (332) has shown that correlations between these properties do exist. Sampling is very quick and cheap and the test may be particularly valuable in checking where high temperatures are suspected but there is no visible damage. The depth of heat penetration into a member can also be conveniently monitored by taking samples at prescribed depths. The method is still at the development stage but does appear to offer considerable potential for the detailed investigation of the extent and severity of fire damage.

9.13 Specialized instrumental methods

A number of highly specialized laboratory techniques are available to the cement chemist. These require expensive equipment and considerable experience, but may be used to identify various characteristics of the hardened concrete. Some of those which are more commonly used are outlined below.

9.13.1 X-ray fluorescence spectroscopy

A sample of concrete is bombarded by high energy X-rays and the fluorescent emission spectrum so caused is collimated into a parallel beam, directed on to the analysing crystal within a spectrometer and reflected into a detector. The wavelengths and densities of the fluorescent emission are measured and the constituent elements, together with their properties, can be calculated from these data.

The sample of concrete must be in the form of a pellet of suitable density, formed by compressing a dried, finely-ground sample together with a binder under very high pressure. The preparation of such a sample (40 mm in diameter) by compaction of 10 g concrete for 10 seconds by a load of 20 tonnes is described in reference 315, for use in the determination of cement content and chloride content. Actual analysis time is very short, and samples may be reused as required. The method is comparative — the emission results are compared with samples of known properties in terms of the component under investigation.

9.13.2 Differential thermal methods

Differential thermal analysis (DTA) is the best known of these methods, because of its important role in the assessment of HAC concrete. This method involves heating a small sample of powdered concrete in a furnace together with a similar sample of inert material. The rate of temperature rise of the inert sample is controlled to be as nearly uniform as possible, and is measured by thermocouple. The test sample is similarly monitored to provide a trace of the temperature difference between the two specimens, and this trace will have a series of peaks at particular temperatures which are characteristic for the minerals in the sample. These correspond to the loss of water of crystallization of the various mineral forms, and in general form will identify the presence of particular minerals. The method may also be made quantitative by calibrating the apparatus against suitable pure minerals to relate peak height to mass.

DTA, together with the similar methods of differential scanning calorimetry (DSC) and derivative thermogravimetry (DTG) has been described by Midgley (333) and used for following the hydration processes of high alumina concretes. Midgley and Midgley have also described their application to measuring the degree of conversion of HAC concrete (334). Both DTA and DSC have become established techniques for this purpose. A typical DTA trace is shown in Figure 9.7. As the relative proportions of the hydration products change with time due to conversion, so do the heights of the corresponding endotherms, enabling the composition of the matrix and degree of conversion to be determined. Although primarily used to establish degree of conversion, the method will also indicate the presence of sulphate attack or other cement type. Particular care is required in sampling to avoid heating of the sample and resulting moisture loss from the contained materials, which will reduce the reliability of results. Further problems can occur if clay minerals or mica are present in the aggregate. A powdered sample will normally be obtained by drilling the in-situ concrete using a rotary percussion

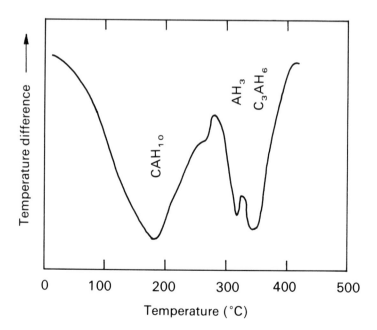

Figure 9.7 Typical differential thermal analysis trace (based on ref. 334).

drill, although a larger sample, such as a core, may also be ground and used. In experienced hands the method is reliable, and an accuracy of $\pm 5\%$ is possible for HAC percentage conversion measurements. Whilst HAC applications have made this method known to many engineers, other applications are also possible, although they are likely to be primarily of a research nature.

9.13.3 *Thermogravimetry, X-ray diffraction, infrared and atomic absorption spectrometry and scanning electron microscopy*

These are all highly specialized techniques which may be used for analysis of the constituents of hardened concrete. Although involving high capital expenditure, they are likely to permit more rapid throughput than chemical approaches and to be able to deal with more complex problems. Many testing organizations have developed their own expertise and procedures with specific methods, and use is growing steadily, often in conjunction with classical techniques. Descriptions of the techniques and applications to the testing of cement and concrete have been given by Ramachandran (335). If, however, precise details of the testing procedures are required, reference should be made to texts such as that by Willard *et al.* (336).

Appendix A: Typical cases of test planning and interpretation of results

The engineer has complete and absolute authority as to whether concrete is condemned or accepted. The problem of testing, and interpretation of the results, will however be approached in a variety of ways — specifications, which will be used as the basis for decisions, vary widely and in some cases may legally empower the engineer to condemn concrete if the cubes fail, irrespective of the condition or quality of the in-situ concrete.

Many factors can however vitiate cube results including variations due to failure to observe the required standardized procedures for sampling, manufacture and curing of the cubes. Further errors may also be introduced by the testing operative or inaccuracies in the testing machine, although these should be checked by regular comparative reference testing. Whilst testing of the in-situ concrete eliminates most of these sources of error, specifications rarely mention in-situ strength and Codes of Practice do not define the in-situ strength required. BS 8110 (320), however, implies that an in-place strength of $f_{cu}/1.5$ is expected for in-situ work by the adoption of a partial factor of safety of 1.5 when calculating the design concrete strength to use in calculations. If a design is based on some other Code of Practice this value may vary, but the basic principle that in-situ strength is recognized as being lower than standard cube specimen strength remains. It is to be hoped that engineers facing the problem of failed cubes will consider these aspects, as well as the in-situ requirements of the concrete and possible errors in the cube results, before taking decisions.

The following examples are included to assist the planning of tests and interpretation of results for this and other commonly occurring situations.

A1 28-day cubes fail (cube results suspect)

(i) *Problem*
28-day cubes from concrete used for in-situ beam construction are of low strength. Visual examination of the cubes suggests that they were poorly made.

(ii) *Aims of testing*
The principal aim will be confirmation of specification compliance of the in-situ concrete, followed by determination of structural adequacy if non-compliance is indicated.

(iii) *Proposals*

Schmidt hammer and/or ultrasonic pulse velocity testing of the beams is suggested, either by way of comparison with similar members of known acceptable standard cube strength, or to yield an absolute in-situ strength prediction based on a specifically prepared calibration for the particular mix. Visual comparison may also prove valuable. Tests should be located at mid-depth or spread evenly to provide a representative average value of in-situ strength for comparison with the specified value after application of safety factors. If doubt still exists then cores can be used, located to give representative values from the suspect concrete.

(iv) *Interpretation*

If similar members are available for comparison with those that are suspect, the combined raw non-destructive test results should be plotted in histogram form and the overall coefficient of variation calculated to indicate the uniformity of concrete between members. The mean values for each group should also be calculated for comparison, bearing in mind possible minor variations due to age differences. If non-uniformity is indicated, relative strengths may be estimated from the mean non-destructive results for each group.

   If similar satisfactory members are not available, or comparative non-destructive testing suggests borderline values, strength calibration charts for the particular mix may be used. If these are not available, or cannot be obtained, then cores cut to examine representative concrete of the suspect members should be used to estimate the equivalent 'standard' cube strength (or potential cube strength).

(v) *Numerical example*

Specified 28-day characteristic cube strength $= 30 \, \text{N/mm}^2$. Schmidt hammer and UPV comparisons with similar beams cast one week earlier show only one peak in histogram form, and have the following values.

Schmidt hammer: mean rebound no. $= 32$
                coefficient of variation $= 6\%$

UPV: mean velocity $= 4.15 \, \text{km/sec}$
                coefficient of variation $= 4\%$.

Mean standard 28-day cube strengths for comparison beams $= 34 \, \text{N/mm}^2$,

hence

Estimated mean standard 28-day cube strength for suspect beams

        $= 32 \, \text{N/mm}^2$ based on Schmidt hammer results
        $= 35 \, \text{N/mm}^2$ based on ultrasonic pulse velocities

(using calibration curves of standard form).

Whilst NDT results suggest only one supply and reasonable construction quality, the estimated mean equivalent standard 28-day cube strength for the suspect concrete is low in relation to the expected value of $30 + 1.64s$

where $s = 5\,\text{N/mm}^2$ (corresponding to 'normal' standards—Table 1.8)

giving $38\,\text{N/mm}^2$ in-situ strength.

The effects of age difference on these values can be assumed to lead to a small underestimate of true strength, but this cannot be relied upon. It appears therefore that whilst the mean value of strength for the suspect batches is above the minimum specified, the proportion of concrete likely to be below this value may be greater than normally permitted.

If further evidence is required cores should be used. These should be taken near mid-depth of a typical beam and at least four should be used to provide an estimate of potential strength as described in Appendix C. This estimate, which will have at best an accuracy of $\pm 15\%$, can then be compared with the absolute minimum specified ($0.85 \times 30 = 25.5\,\text{N/mm}^2$ for BS 8110).

Core results

estimated potential strength = 27
29
32
35
Mean = $\overline{31}\,\text{N/mm}^2 \pm 4.5\,\text{N/mm}^2$.

Since the mean is above the characteristic specified value, and all results exceed the minimum acceptable value, it cannot be proved conclusively that the concrete does not meet the specification. Such results could well form part of an acceptable spread of values within a characteristic strength of $30\,\text{N/mm}^2$ and the concrete should not be rejected.

A2   28-day cubes fail (cube results genuine)

(i) *Problem*
A suspended floor slab has been cast from several batches and the 28-day cube strengths (6 cubes) are below the characteristic value required. There is no reason to suspect the validity of the cube results.

(ii) *Aims of testing*
The initial aim will be to establish whether the low cubes are representative or relate to isolated substandard batches. The subsequent aim will then be to assess structural adequacy.

(iii) *Proposals*
Visual inspection may indicate uniformity or otherwise of the slab in the first instance. This can be followed by Schmidt hammer tests on the soffit to confirm uniformity. Direct ultrasonic tests may prove difficult, and with indirect readings on the soffit being unreliable cores should be taken from typical zones. In cases of doubt, in-situ load testing may be necessary.

(iv) *Interpretation*
Schmidt hammer readings should be taken on a regular grid, and plotted on a 'contour' plan to indicate the degree of uniformity. A histogram plot may also be worth while to provide confirmation of this. The results from cores can be used to estimate an average actual in-situ strength which can then be related to the required design strength with an allowance for the likely standard deviation of in-situ results.

(v) *Numerical example*
Specified characteristic 28-day cube strength $= (f_{cu}) = 40 \, \text{N/mm}^2$
Mean 'standard' 28-day cube strength $= 38 \, \text{N/mm}^2 \pm 2 \, \text{N/mm}^2$
Mix design mean strength $= 46 \, \text{N/mm}^2$.

Visual inspection and non-destructive test results suggest uniform construction across the slab.

Average estimated in-situ cube strength from six cores $= 30 \pm 1.5 \, \text{N/mm}^2$.

Estimated characteristic in-situ strength range $= \begin{cases} 31.5 - 1.64s' \, \text{N/mm}^2 \\ 28.5 - 1.64s' \, \text{N/mm}^2. \end{cases}$

Adopt estimated in-situ standard deviation $(s') = 4.5 \, \text{N/mm}^2$

based on scatter of Schmidt hammer results and past site cube records which shows 'good' control (see Table 1.8).
    Hence

estimated characteristic in-situ strength $= \begin{cases} 31.5 - 7.5 = 24 \, \text{N/mm}^2 \\ 28.5 - 7.5 = 21 \, \text{N/mm}^2. \end{cases}$

If the design is to BS 8110 (320), the partial factor of safety on concrete strength is 1.5, hence minimum acceptable design strength $= 40/1.5 = 26 \, \text{N/mm}^2$. Since the estimated range of characteristic in-situ strength lies below the minimum design strength, the concrete must be considered unacceptable. If the slab is to be permanently dry, the estimated in-situ values may be increased by 10%, but this factor will not be sufficient to accept the concrete unless the slab is not critically stressed. It is recommended that an in-situ load test be undertaken to establish directly the serviceability behaviour of the slab. Likely durability performance must also be considered based on exposure conditions.

A3   Cubes non-existent for new structure

(i) *Problem*
A large number of columns have been cast, and the cubes lost.

(ii) *Aims of testing*
The principal aim will generally be confirmation of acceptability of the in-situ concrete from the point of view of strength and durability.

(iii) *Proposals*
A comprehensive comparative survey, using surface hardness and/or ultrasonic pulse velocity. Visual inspection may also indicate lack of uniformity. Plot raw results to indicate patterns and then follow up with a limited number of cores at points of apparently lowest and highest strength, unless reliable calibrations for the mix are available or can be obtained. If similar members are available of concrete which is known to be acceptable it may be adequate to rely on non-destructive comparisons with these, using cores only in cases of extreme doubt. Reserve crushed cores for chemical analysis to determine the cement content if strengths indicate durability doubts.

(iv) *Interpretation*
It is important that the non-destructive results to be used comparatively are taken at comparable points in relation to the members tested. This is because of the likely within-member strength variations, and measurements should preferably be taken at points of expected lower strength (i.e. near the top of the columns). Sufficient readings should be taken to encompass the various batches of concrete that may have been used.

Results should be plotted in histogram form to detect the weakest areas; coefficients of variation may also provide valuable confirmation of construction uniformity. The cores should provide values for minimum strength to be compared with a calculated minimum acceptable value from the design, as well as a rough calibration for the non-destructive tests. If attempts are to be made to relate results to specifications, the likely within-member variations and in-situ/standard specimen strength must not be overlooked. Under normal circumstances a minimum in-situ strength of characteristic/1.5 would be acceptable for design to BS 8110 (320), whilst an even lower value may be adequate for low stress areas, subject to adequate durability as indicated by cement content.

(v) *Numerical example*
Specified characteristic cube strength $(f_{cu}) = 30\,\text{N/mm}^2$.
For design to BS 8110 (320) assume

$$\text{minimum acceptable in-situ strength} = \frac{f_{cu}}{1.5} = 20\,\text{N/mm}^2.$$

For four cores taken to correspond with lowest pulse velocities and rebound numbers:

estimated in-situ cube strengths 20.5 N/mm²
                                 25.0 N/mm²
                                 22.5 N/mm²
                                 21.0 N/mm²
                         Mean 22.0 N/mm².

∴ Estimated minimum in-situ strength = 22 N/mm² ± 1.5 N/mm².

Hence the mean and all results are above minimum acceptable value, and the concrete will be considered adequate. It will follow that the remainder of the concrete is also acceptable since these results relate to the worst locations.

Note (1): If either individual results or the mean estimated in-situ strength are below the minimum acceptable as calculated above (based on BS 8110) detailed consideration should be given to design stress levels, and service moisture conditions.

Note (2): If the concrete strength is critical to the design, or if calibrations have been used with surface hardness or UPV results to estimate strength without cores, it may be appropriate to include a factor of safety to account for this, e.g. maximum acceptable design stress = 22/1.2 = 18 N/mm² based on the factor of safety of 1.2 recommended by BS 6089 (12).

A4   Cubes damaged for new structure

(i) *Problem*
A series of elements have been cast but the cubes damaged. The scope for in-situ testing is very limited due to access difficulties.

(ii) *Aims of testing*
The principal aim is to estimate the in-situ characteristic strength.

(iii) *Proposals*
Use statistical procedures to establish the likely in-situ characteristic strength based on tests at six accessible locations.

(iv) *Interpretation*
For the six results obtained, the mean estimated in-situ strength is 36 N/mm² and the coefficient of variation of these results is 12%.

Using the 'normal' distribution relationship illustrated in Figure 1.10, it can be seen by interpolation that for six values, the estimated in-situ characteristic strength with 95% confidence limits, is

$$f_{cu} \simeq 0.7 f_c$$
$$= 0.7 \times 36 = 25 \, \text{N/mm}^2.$$

A5 Cubes non-existent for existing structure

(i) Aims of testing
The aim of testing will be to provide a concrete strength estimate for use in design calculations relating to a proposed modification of the structure.

(ii) Proposals
Survey by ultrasonic pulse velocity, Capo, pull-off or Windsor probe, correlated with a limited number of cores according to practical limitations. Tests to be spread as representatively as possible over members under examination.

(iii) Interpretation
Use mean estimated in-situ cube strength to obtain a value of design strength which takes account of the standard of construction quality as well as uncertainties about the adequacy of the in-situ test data.

(iv) Numerical example
Estimated mean in-situ cube strength $= 25 \, \text{N/mm}^2$
Assumed mean 'standard' cube strength $= 25 \times 1.5 = 37.5 \, \text{N/mm}^2$

for 'normal' construction quality (unless evidence from scatter of test results suggests otherwise), standard deviation of control cubes estimated at $5 \, \text{N/mm}^2$ (Table 1.8).

Hence estimated characteristic 'standard' cube strength $= 37.5 - 1.64 \times 5$
$$= 29 \, \text{N/mm}^2.$$

Allowance should be made for errors in test data by a factor of safety (1.2 suggested by BS 6089 (12)).

Hence maximum design stress using a partial factor of safety of 1.5 on concrete strength for design to BS 8110 (320) equals:

$$\frac{29}{1.2 \times 1.5} = 16 \, \text{N/mm}^2.$$

Alternatively, if the test results are taken to correspond to locations of lowest anticipated strength:

Maximum design stress = estimated minimum in-situ cube strength, given by mean of test results/1.2.

A6 Surface cracking

(i) *Problem*
The wing wall to a highway bridge abutment shows random surface cracks and spalling several years after construction.

(ii) *Aims of testing*
The principal aim will be identification of the cause of deterioration followed by an assessment of present and future serviceability. Appointment of blame may follow.

(iii) *Proposals*
Visual inspection of crack patterns, and their development with time, may permit preliminary classification of cause as (a) structural actions, (b) shrinkage, or (c) material deterioration. This may be followed by strength assessment as in A2 if structural actions are suspected, or chemical/petrographic testing if material deterioration is likely. Cores may conveniently be used to provide suitable samples and should be taken from the areas most seriously affected. Chemical testing to detect chlorides or sulphates will be selected according to the crack pattern whilst microscopic examination can check for frost action, alkali/aggregate reaction and entrained air content.

Serviceability will be determined on the basis of the extent of deterioration and the ability to prevent worsening of the situation.

(iv) *Interpretation*
Reference to Table 1.2 will assist preliminary identification. Shrinkage cracks are likely to occur at an early age and follow a recognizable pattern, as do cracks due to structural actions. Material deterioration is therefore indicated in this case, and may be due to chemical attack from internal or external sources or due to frost action. Chloride attack is unlikely since the cracks do not follow the pattern of reinforcement, thus initially test for sulphate and cement content. If the results of these tests indicate acceptable levels, petrographic examination will be necessary to attempt to identify aggregate/alkali attack or frost action. If frost action is indicated, micrometric examination will yield an estimate of the entrained air content for comparison with the specified value. Expansion and alkali-immersion tests on cores may be required if alkali/aggregate reaction is found.

If future deterioration can be prevented by protection of the concrete from the source of attack, this should be implemented after such cutting out and making good as may be necessary. If the source of deterioration is internal and not of a localized nature it may prove necessary to replace the member once it reaches a condition of being unfit for use.

A7   Reinforcement corrosion

(i) *Problem*
A major modern in-situ concrete structure is showing numerous rust-stained cracks, and in some cases pieces of concrete have spalled exposing seriously corroded reinforcement. Repairs are obviously necessary to prevent continued deterioration and to restore the appearance of the structure.

(ii) *Aims of testing*
The principal aim will be to establish the reason for the corrosion as well as its (present and likely future) extent (which may not be visible) to permit the design of repair proposals. Litigation may follow. Structural adequacy must also be checked.

(iii) *Proposals*
A comprehensive covermeter survey coupled with phenolphthalein carbonation depth measurements should be undertaken. If access costs are high, this may be limited initially to about 25% of the areas involved although a full survey will be necessary prior to repair. A limited number of chloride tests at the level of the steel using simple methods (Hach or Quantab) should indicate acceptable or excessive chloride levels. A limited number of cores should also be taken for compressive strength testing. If carbonation is found to be excessive these should also be tested for absorption and cement content. Further cores or more detailed chemical analysis may be required (chloride, cement content and possibly water/cement ratio) depending upon results for litigation purposes.

A comparative half-cell potential survey may be useful in determining the corrosion risk in apparently undamaged regions unless this can be established from the covermeter and carbonation survey. ISAT measurements may also indicate the extent of highly absorptive concrete if that is found to be present.

(iv) *Interpretation*
The most likely causes of the problem are the existence of inadequate cover and/or excessive carbonation. These will be immediately apparent from the covermeter and phenolphthalein survey. The reasons for excessive carbonation may be deducible from the tests on cores, and may include mix deficiencies indicated by low (or variable) strength or low cement content, and inadequate curing possibly indicated by high water absorption. Core strength results coupled with careful inspection of corroded steel should enable structural adequacy to be checked, and the need for additional steel or strengthening to be determined. If chloride content is found to be high, more extensive testing to provide surface zone gradients may help to identify likely sources. A half-cell potential survey may be useful in this situation to help plan the repair strategy.

The information provided by these tests may be adequate to permit the design of repairs, but deterioration is often the result of a combination of several factors and it is important that all reasonable possibilities are considered before conclusions are reached.

Appendix B: Examples of pulse velocity corrections for reinforcement

See section 3.3.2.5 for full details of notation and procedures.

Example B1

The beam contains 16 mm diameter main bars with 10 mm links as shown in Figure B1. The measured pulse velocity across the top of the beam away from links is 4.2 km/s. Determine the anticipated measured apparent velocity when readings are made directly in line with a link.

As main bar will have no practical influence on measured values, $V_c = 4.2$ km/s. Hence from Figure 3.15 for $V_c = 4.2$ km/s and 10 mm bar,

$$\gamma = 0.88$$

thus allowing for 25 mm cover at each end of link

$$\frac{L_s}{L} = \frac{150}{200} = 0.75$$

thus

$$k = 1 - 0.75(1 - 0.88)$$
$$= 0.91$$

$$\therefore \text{ anticipated } V_m = \frac{4.2}{0.91} = 4.62 \text{ km/s}.$$

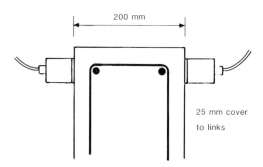

200 mm

25 mm cover
to links

Figure B1

Example B2

An uncracked beam shown in Figure B2 has well bonded 20 mm bars with 50 mm end cover. Measurements at 150 mm offset yield $V_m = 4.65$ km/s. Estimate the true pulse velocity in the concrete (V_c).

$$a > 2b \quad \text{and} \quad a/L = 0.075.$$

For trial values of V_c evaluate kV_m as in Table B1, and plot as shown in Figure B3.
    On the basis of these values, try $V_c = 4.10$ then $\gamma = 0.79$, $k = 0.88$ hence

$$V_m = \frac{4.10}{0.88} = 4.66 \text{ km/s (OK)}$$

thus

$$\text{estimated } V_c = 4.10 \text{ km/s}$$

(note: RILEM 'maximum effect' factors suggest $V_c = 3.75$ km/s).

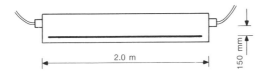

Figure B2

Table B1

Trial V_c	γ (Fig. 3.15)	k (Fig. 3.16)	kV_m	$(V_c - kV_m)$
4.5 km/s	0.87	0.94	4.37 km/s	+0.13
3.5 km/s	0.68	0.79	3.67 km/s	−0.17

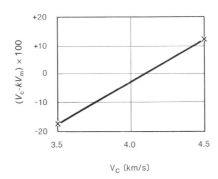

Figure B3

Example B3

Measurements across a 300 mm wide beam containing three No. 32 mm bars give a value of $V_m = 4.4$ km/s. Estimate V_c.

Then
$$\frac{L_s}{L} = \frac{3 \times 32}{300} = 0.32.$$

For trial values of V_c, evaluate kV_m as in Table B2 and plot Figure B4. Thus try $V_c = 4.25$ km/s, then $\gamma = 0.89$ and $k = 0.965$ hence

$$V_m = \frac{4.25}{0.965} = 4.40 \text{ km/s (OK)}$$

thus estimated $V_c = 4.25$ km/s (note RILEM 'maximum effect' factors suggest $V_c \simeq 3.9$ km/s).

Table B2

Trial V_c	γ (Fig. 3.18)	$k\left(=1-\dfrac{L_s}{L}(1-\gamma)\right)$	kV_m	$(V_c - kV_m)$
4.5 km/s	0.92	0.974	4.28 km/s	+0.22
4.0 km/s	0.85	0.952	4.19 km/s	−0.19

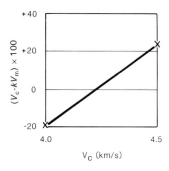

Figure B4

Appendix C: Example of evaluation of core results

A 100 mm diameter core drilled horizontally from a wall of concrete with 20 mm maximum aggregate size contains one no. 20 mm reinforcing bar normal to the core axis and located at 35 mm from one end.
Measured water-soaked concrete density = 2320 kg/m³ after correction for included reinforcement (section 5.1.3.4).
Measured crushing force = 160 kN (following BS 1881 (114) procedure).
Failure mode — normal.
Measured core length after capping = 120 mm.

(i) *BS 1881 (114) in-situ cube strength*

$$\text{Measured core strength} = \frac{160 \times 10^3}{\pi \times \dfrac{100^2}{4}} = 20.5 \text{ N/mm}^2.$$

Core length/diameter ratio = 120/100 = 1.2.

$$\text{Estimated in-situ cube strength} = \frac{2.5}{\left(1.5 + \dfrac{1}{1.2}\right)} \times 20.5 \text{ for a horizontal core}$$

$$= 22 \text{ N/mm}^2.$$

$$\text{Reinforcement correction factor} = 1 + 1.5\left(\frac{20}{100} \times \frac{35}{120}\right)$$

$$= 1.09.$$

Corrected in-situ cube strength $= 22 \times 1.09 \text{ N/mm}^2 \pm 12\%$ for an individual result.

$$= 24.0 \pm 3 \text{ N/mm}^2.$$

(ii) *Concrete Society (25) potential cube strength*
Measured core strength = 20.5 N/mm².

Core length/diameter ratio = 1.2.

$$\text{Estimated potential cube strength} = \frac{3.25}{\left(1.5 + \dfrac{1}{1.2}\right)} \times 20.5 \text{ for a horizontal core}$$

$$= 28.5 \text{ N/mm}^2.$$

Reinforcement correction factor $= 1.09$.
Potential density of concrete $= 2350 \text{ kg/m}^3$ (mean value from cubes).

$$\text{Excess voidage} = \left(\frac{2350 - 2320}{2350 - 500} \right) \times 100\%$$

$$= 1.6\%.$$

Strength multiplying factor $= 1.14$ (Figure 5.4).
Corrected potential cube strength $= 28.5 \times 1.09 \times 1.14$
$$= 35.5 \text{ N/mm}^2.$$

(*Note*: Accuracy cannot be realistically quoted for a single result but the mean estimated potential cube strength from a group of at least four cores may be quoted to between $\pm 15\%$ and $\pm 30\%$ subject to a procedure described by Technical Report No. 11 (25) to eliminate abnormally low results. In this instance because of the use of reinforcement and excess voidage correction factors, the estimated accuracy for a group of four is unlikely to be as high as $\pm 15\%$.)

(iii) *Procedure to eliminate abnormal results* (*Reference 25*)
This requires that for n cores (where n is at least 4) the lowest value is separated and the value t calculated from

$$t = \frac{\text{mean of remainder} - \text{lowest}}{\dfrac{\text{mean of remainder} \times 6}{100} \times \sqrt{1 + \dfrac{1}{n - 1}}}.$$

The value t is then compared with Table C1, and if greater than the value in column A, corresponding to the total number of cores n, the lowest core result is discarded if there is any evidence of abnormality in relation to the others.

This may be in terms of location, reinforcement, compaction, cracks or drilling damage. If the value of t is greater than that in column B, the lowest result is discarded irrespective of other considerations. The mean value

Table C1

No. of cores (n)	t	
	A	B
4	2.9	4.3
5	2.4	3.2
6	2.1	2.8
7	2.0	2.6
8	1.9	2.5

obtained from the remaining $(n - 1)$ cores is then taken as the estimated potential strength.

If an abnormally high result is obtained, although this is less likely, the same procedure can be adopted but substituting the highest value for the lowest. Applying this procedure to the results used in section A1 (p. 256), i.e. four cores with potential strengths 27, 29, 32, 35 N/mm²,

$$t = \frac{32 - 27}{\dfrac{32 \times 6}{100}\sqrt{1 + \tfrac{1}{3}}}$$

$$= 2.26.$$

This is less than the value 2.9 from column A, Table C1, and the quoted mean and accuracy may be considered valid.

References

(1) Petersen, C.G. and Poulsen, E. *Pull-out testing by Lok-test and Capo-test*. Dansk Betoninstitut A/S, 1992.
(2) Somerville, G. The design life of concrete structures. *Structural Engineer*, **64A**, No. 2, Feb. 1986, 60–71.
(3) Clifton, J.R. Predicting the service life of concrete. *ACI Materials Journal*, **90**, No. 6, Nov./Dec. 1993, 611–617.
(4) *Inspection and Maintenance of Reinforced and Prestressed Concrete Structures*. Federation Internationale de Pre-contrainte (FIP), Thomas Telford Ltd, London, 1986.
(5) *Appraisal of existing structures*. Institution of Structural Engineers, London, 1980.
(6) RILEM. Draft recommendation for damage classification of concrete structures. *Materials and Structures*, **27**, 1994, 362–369.
(7) ACI Committee 364. Guide for evaluation of concrete structures prior to rehabilitation. *ACI Materials Journal*, **90**, No. 5, Sept./Oct. 1993, 479–498.
(8) Currie, R.J. and Crammond, N.J. Assessment of existing high alumina cement construction in the UK. *Proc. ICE, Structs & Bldgs*, **104**, Feb. 1994, 83–92.
(9) Assessment and repair of fire-damaged concrete structures. *Tech. Rept.* **33**, Concrete Society, London, 1990.
(10) Fleischer, C.C. and Chapman-Andrews, J.F. Survey and analysis of bomb damaged reinforced concrete structures. *Proc. Struct. Faults and Repairs 93*, Eng. Technics Press, Edinburgh, **2**, 1993, 111–118.
(11) BS 1881: Part 201 *Guide to the use of non-destructive methods of test for hardened concrete*. British Standards Institution, London.
(12) BS 6089 *Guide to assessment of concrete strength in existing structures*. British Standards Institution, London.
(13) Bungey, J.H. Testing concrete in structures – a guide to equipment for testing concrete in structures. *TN* **143**, CIRIA, 1992, 87p.
(14) Shickert, G. NDT in civil engineering in Germany. *Insight*, **36**, No. 7, July 1994, 489–495.
(15) Carino, N.J. Nondestructive testing of concrete: History and challenges. *Spec. Publ.* **SP 144-30**, American Concrete Institute, Detroit, 1994, 623–678.
(16) DeBreto, J., Branco, F.A. and Ibanez, M. A knowledge based concrete bridge inspection system. *Concrete International*, **16**, No. 2, Feb. 1994, 59–63.
(17) Bungey, J.H. The most effective use of NDT of concrete. *Proc. 4th European Conf. on NDT*, Vol. 2. Pergamon, Oxford, 1988, pp. 963–970.
(18) Smith, J.R. Assessment of concrete buildings. *Concrete International*, **13**, No. 12, Dec.. 1991, 48–49.
(19) Pollock, D.J., Kay, E.A. and Fookes, P.G. Crack mapping for investigation of Middle East concrete. *Concrete*, **15**, No. 5, May 1981, 12–18.
(20) Non-structural cracks in concrete. *Tech. Rept.* **22**, Concrete Society, London, 1992.
(21) Higgins, D.D. Diagnosing the causes of defects or deterioration in concrete structures. Current Practice Sheet 69, *Concrete*, **15**, No. 10, Oct. 1981, 33–34.
(22) Fewtrell, A. Special access techniques for inspection works. *Proc. Struct. Faults and Repairs 93*, Eng. Technics Press, Edinburgh, **2**, 1993, pp. 119–124.
(23) Palmer, D. *The diagnosis of alkali–silica reaction*. Report of Working Party, British Cement Association, Slough, 1988.
(24) Bungey, J.H. and Madandoust, R. Evaluation of non-destructive strength testing of lightweight concrete. *Proc. ICE, Structs & Bldgs*, **104**, Aug. 1994, 275–283.
(25) Concrete core testing for strength. *Tech. Rept.* **11**, Concrete Society, London, 1987.
(26) Bungey, J.H. Concrete strength variations and in-place testing. *Proc. 2nd Australian Conf. on Engineering Materials*, University of New South Wales, Sydney, 1981, 85–96.

(27) Bungey, J.H. and Madandoust, R. Strength variations in lightweight concrete beams. *Cement and Concrete Composites*, **16**, 1994, 49–55.

(28) Maynard, D.P. and Davis, S.G. The strength of in-situ concrete. *Structural Engineer*, **52**, No. 10, Oct. 1974, 369–374.

(29) Davies, S.G. Further investigations into the strength of concrete in structures. *Tech. Rept.* **42.514**, Cement and Concrete Association, Slough, 1976.

(30) Miao, B. *et al.* Influence of concrete strength on in-situ properties of large columns. *ACI Materials Journal*, **90**, No. 3, May/June 1993, 214–219.

(31) Murray, A.McC. and Long, A.E. A study of the in-situ variability of concrete using the pull-off method, *Proc. ICE*, Part 2, **83**, Dec. 1987, 731–745.

(32) Tomsett, H.N. The development of a nondestructive test analysis. *Proc. Nondestructive Testing in Civil Engineering*, Brit Inst NDT, **2**, 1993, 435–450.

(33) Tomsett, H.N. Ultrasonic pulse velocity measurements in the assessment of concrete quality. *Magazine of Concrete Research*, **32**, No. 110, March 1980, 7–16.

(34) Leshchinsky, A.M., Leshchinsky, M.Yu. and Goncharova, A.S. Within-test variability of some non-destructive methods for concrete strength determination. *Magazine of Concrete Research*, **42**, No. 153, Dec.. 1990, 245–248.

(35) Hindo, K.R. and Bergstrom, W.R. Statistical evaluation of the in-place strength of concrete. *Concrete International*, **7**, No. 2, Feb. 1985, 44–48.

(36) Stone, W.C., Carino, N.J. and Reeve, C.P. Statistical methods for in-place strength predictions by the pull-out test. *ACI Journal*, **83**, No. 5, Sept./Oct. 1986, 745–756.

(37) ACI Committee 228. In-place methods for determination of strength of concrete. ACI 228, **IR-89**. American Concrete Institute, Detroit, 1989.

(38) Carino, N.J. Statistical methods to evaluate in-place test results. *Proc. Int. Workshop on testing during Concrete Construction (RILEM)*, ed. Reinhardt, H.W., Chapman & Hall, 1992, pp. 432–448.

(39) Leshchinsky, A.M. Non-destructive testing of concrete strength: statistical control. *Materials and Structures*, **25**, No. 146, March 1992, 70–78.

(40) Bartlett, F.M. and MacGregor, J.G. Equivalent specified concrete strength from core test data. *Concrete International*, **17**, No. 3, March 1995, 52–58.

(41) Reynolds, W.N. Measuring concrete quality non-destructively. *British Journal of NDT*, **26**, No. 1, Jan. 1984, 11–14.

(42) Samarin, A. *Combined Methods. Handbook on Non-destructive Testing of Concrete*, ed. Malhotra, V.M. and Carino, N., CRC Press, Ch. 8, 1991, pp. 189–201.

(43) RILEM. Draft recommendation for in-situ concrete strength determination by combined non-destructive methods. *Materials and Structures*, **26**, 1993, 43–49.

(44) Galan, A. Estimate of concrete strength by ultrasonic pulse velocity and damping constant. *ACI Journal*, **64**, No. 10, Oct. 1967, 678–684.

(45) BS 1881: Part 202 *Recommendations for surface hardness testing by rebound hammer*. British Standards Institution, London.

(46) ASTM C805 *Rebound number of hardened concrete*. American Society for Testing and Materials, Philadelphia.

(47) Akashi, T. and Amasaki, S. Study of the stress waves in the plunger of a rebound hammer at the time of impact. *Spec. Publ.* **SP 82-2**, American Concrete Institute, Detroit, 1984, 17–34.

(48) Kolek, J. Non-destructive testing of concrete by hardness methods. In *Non-destructive Testing of Concrete and Timber*. Institution of Civil Engineers, London, 1970, 19–22.

(49) Willetts, C.H. Investigation of the Schmidt concrete test hammer. *Misc. Papers*, S-627, US Army Engineer Waterways Experiment Station, Vicksburg, Miss., June 1958.

(50) Malhotra, V.M. *Testing hardened concrete: non-destructive methods*. Monograph 9, American Concrete Institute, Detroit, 1976.

(51) BS 1881: Part 116 *Method for determination of compressive strength of concrete cubes*. British Standards Institution, London.

(52) Chaplin, R.G. Abrasion resistant concrete floors. In *Advances in Concrete Slab Technology*, Pergamon, Oxford, 1980, pp. 532–543.

(53) Andrews, D.R. Future prospects for ultrasonic inspection of concrete. *Proc. ICE Structs & Bldgs*, **99**, Feb. 1993, 71–73.

(54) Bungey, J.H. Ultrasonic testing to identify alkali–silica reaction in concrete. *Brit. J. NDT*, **33**, No. 5, 1991, 227–231.

(55) Hillger, W. Imaging of defects in concrete by ultrasonic pulse-echo technique. *Proc. Struct. Faults and Repairs 93*, Eng. Technics Press, Edinburgh, **3**, 1993, pp. 59–65.

(56) Kroggel, O. Ultrasonic examination of crack structures in concrete slabs. *Proc. Struct. Faults and Repairs 93*, Eng. Technics Press, Edinburgh, **3**, 1993, pp. 67–70.

(57) Sack, D.A. and Olson, L.D. Advanced NDT methods for evaluating concrete bridges and other structures. *Proc. Struct. Faults and Repairs 93*, Eng. Technics Press, Edinburgh, **1**, 1993, pp. 65–76.

(58) BS 1881: Part 203 *Recommendations for measurement of velocity of ultrasonic pulses in concrete*. British Standards Institution, London.

(59) ASTM C597 *Standard test method for pulse velocity through concrete*. American Society for Testing and Materials. Philadelphia.

(60) *Instruction manual for model C-4899 V-meter*. James Electronics Inc., Chicago, Illinois.

(61) PUNDIT, CNS Electronics Ltd, 61–63 Holmes Rd, London.

(62) Jones, R. *Non-destructive Testing of Concrete*. Cambridge University Press, London, 1962.

(63) Ikpong, A.A. The relationship between the strength and non-destructive parameters of rice husk ash concrete. *Cement and Concrete Research*, **23**, 1993, 387–398.

(64) Jones, R. and Facaoaru, I. Recommendations for testing concrete by the ultrasonic pulse method. *Materials and Structures*, **2**, No. 10, 1969, 275–284.

(65) Bungey, J.H. The validity of ultrasonic pulse velocity testing of in-place concrete for strength. *NDT International*, IPC Press, Dec. 1980, pp. 296–300.

(66) Swamy, R.N. and Al-Hamed, A.H. The use of pulse velocity measurements to estimate strength of air dried cubes and hence in-situ strength of concrete. *Spec. Publ.* **SP 82-13**, American Concrete Institute, Detroit, 1984, 247–276.

(67) Testing of concrete by the ultrasonic pulse method. *NDT* **1**, RILEM, Paris, Dec. 1972.

(68) Bungey, J.H. The influence of reinforcement on pulse velocity testing. *Spec. Publ.* **SP 82-12**, American Concrete Institute, Detroit, 1984, 229–246.

(69) Tomsett, H.N. Non-destructive testing of floor slabs. *Concrete*, **8**, No. 3, March 1974, 41–42.

(70) Amon, J.A. and Snell, L.M. The use of pulse velocity techniques to monitor and evaluate epoxy grout repairs to concrete. *Concrete International*, **1**, No. 12, Dec. 1979, 41–44.

(71) Matti, M.A. Frozen concrete. *Concrete International*, **8**, No. 4, April 1986, 34–41.

(72) Bungey, J.H. Ultrasonic pulse testing of high alumina cement concrete on site. *Concrete*, **8**, No. 9, Sept. 1974, 39–41.

(73) Swamy, R.N. and Laiw, J.C. Evaluation of concrete deterioration due to alkali–silica reactivity through nondestructive tests. *Proc. Nondestructive Testing in Civil Engineering*, Brit Inst NDT, **1**, 1993, 315–328.

(74) Hillger, W. A testing equipment for automatic quality control of precast concrete components. *Proc. Struct. Faults and Repairs 87*, Eng. Technics Press, Edinburgh, 1987, pp. 91–100.

(75) BS 1881: Part 207 *Recommendations for the assessment of concrete strength by near-to-surface tests*. British Standards Institution, London.

(76) Windsor Probe Test System. Technical Data Manual, Sinco Products Inc., PO Box 361, Hog Hill Road, East Hampton, Connecticut CT 06424, USA.

(77) ASTM C803 *Penetration resistance of hardened concrete*. American Society for Testing and Materials, Philadelphia.

(78) Nasser, K.W. and Al-Manaseer, A.A. New non-destructive test. *Concrete International*, **9**, No. 1, Jan. 1987, 41–44.

(79) BS 4078: Part 1 *Powder actuated fixing systems – Code of practice for safe use*. British Standards Institution, London.

(80) Bungey, J.H. Testing by penetration resistance. *Concrete*, **15**, No. 1, Jan. 1981, 30–32.

(81) Nasser, K.W. and Al-Manaseer, A.A. Comparison of non-destructive testers of hardened concrete. *Materials Journal*, American Concrete Institute, Detroit, Sept./Oct. 1987, 374–380.

(82) Swamy, R.N. and Al-Hamed, A.H. Evaluation of the Windsor probe test to assess in-situ concrete strength. *Proc. ICE*, Pt 1, **77**, June 1984, 167–194.

(83) Shoya *et al.* Estimating the compressive strength of concrete from pin-penetration test method. *Proc. Nondestructie Testing in Civil Engineering*, Brit Inst NDT, **1**, 1993, 379–390.

(84) Carino, N.J. *Pull-out test. Handbook on Non-destructive Testing of Concrete*. CRC Press, Boston, USA, Chap. 3, 1991, pp. 39–82.

(85) Germann Aps. Lok-test method for determination of the compressive strength of concrete in place. Lok-test Aps., Emdrupvej 102, DK 2400, Copenhagen.

(86) Petersen, C.G. Lok-test and Capo-test development and their applications. *Proc. ICE*, Pt 1, **76**, May 1984, 539–549.

(87) Bungey, J.H. An appraisal of pull-out methods of testing concrete. *Proc. NDT 83*, Engineering Technics Press, Edinburgh, 1983, pp. 12–21.

(88) Ottosen, N.S. Nonlinear finite element analysis of pull-out test. *Journal of Struct. Division ASCE*, **107**, No. ST4, April 1981, 591–603.

(89) Yener, M. Overview and progressive finite element analysis of pullout tests, *ACI Structural Journal*, **91**, No. 1, Jan./Feb. 1994, 49–58.

(90) Stone, W.C. and Carino, N.J. Deformation and failure in large scale pull-out tests. *ACI Journal*, Nov./Dec. 1983, 501–513.

(91) Stone, W.C. and Giza, B.J. The effect of geometry and aggregate on the reliability of the pull-out test. *Concrete International*, **7**, No. 2, Feb 1985, 27–36.

(92) Bickley, J.A. The evaluation and acceptance of concrete quality by in-place testing. *Spec. Publ.* **SP 82-6**, American Concrete Institute, Detroit, 1984, 95–109.

(93) ASTM C900 *Pull-out strength of hardened concrete*. American Society for Testing and Materials, Philadelphia.

(94) Richards, O. Pull-out strength of concrete. *Spec. Tech. Publ.* **STP 626**, American Society for Testing and Materials, Philadelphia, 1977, 32–40.

(95) Chabowski, A.J. and Bryden-Smith, D.W. A simple pull-out test to assess the strength of in-situ concrete. *Current Paper* **CP 25/77**, Building Research Establishment, Garston, June 1977.

(96) Chabowski, A.J. and Bryden-Smith, D.W. Assessing the strength of in-situ Portland cement concrete by internal fracture tests. *Magazine of Concrete Research*, **32**, No. 11, Sept. 1980, 164–172.

(97) Bungey, J.H. Concrete strength determination by pull-out tests on wedge-anchor bolts. *Proc. ICE*, Pt. 2, **71**, June 1981, 379–394.

(98) Mailhot, G., Bisaillon, A., Carette, G.G. and Malhotra, V.M. In-place concrete strength – new pull-out methods. *ACI Journal*, Dec. 1979, 1267–1282.

(99) Domone, P.L. and Castro, P.F. An expanding sleeve test for in-situ concrete and mortar strength evaluation. *Proc. Struct. Faults and Repairs 87*, Engineering Technics Press, Edinburgh, 1987, pp. 149–156.

(100) Keiller, A.K. Assessing the strength of in-situ concrete. *Concrete International*, **7**, No. 2, Feb. 1985, 15–21.

(101) Long, A.E. and Murray, A.McC. The pull-off partially destructive test for concrete. *Spec. Publ.* **SP 82-17**, American Concrete Institute, Detroit, 1984, 327–350.

(102) Jaegermann, C. A simple pull-out test for in-situ determination of early strength of concrete. *Magazine of Concrete Research*, **41**, No. 149, Dec. 1989, 235–242.

(103) Cleland, D.J. In-situ methods for assessing the quality of concrete repairs. *Proc. ICE, Structs & Bldgs*, **99**, Feb. 1993, 68–70.

(104) McLeish, A. Standard tests for repair materials and coatings–Part 1. Pull-Off Testing. CIRIA, **TN 139**, 1992, 41 p.

(105) Bungey, J.H. and Madandoust, R. Factors influencing pull-off tests on concrete. *Magazine of Concrete Research*, **44**, No. 158, March 1992, 21–30.

(106) Long, A.E., Montgomery, F.R. and Cleland, D. Assessment of concrete strength and durability on site. *Proc. Struct. Faults and Repair 87*, Engineering Technics Press, Edinburgh, 1987, pp. 61–73.

(107) CUR. Determination of the bond strength of mortars on concrete. Centre for Civil Engineering Research and Codes, Gouda, Recommendation 20, 1990.

(108) Johansen, R. In-situ strength evaluation of concrete – the break-off method. *Concrete International*, **1**, No. 9, Sept. 1979, 45–51.

(109) Christiansen, V.T., Thorpe, J.D. and Yener, M. In-situ testing of concrete using break-off method. *Proc. Struct. Faults and Repair 87*, Engineering Technics Press, Edinburgh, 1987, pp. 101–112.

(110) Dahl-Jorgensen, E. and Johansen, R. General and specialized use of the break-off concrete strength testing method. *Spec. Publ.* **SP 82-15**, American Concrete Institute, Detroit, 1984, 293–308.

(111) Carlsson, M., Eeg, I.R. and Jahren, P. Field experience in the use of the break-off tester. *Spec. Publ.* **SP 82-14**, American Concrete Institute, Detroit, 1984, 277–292.

(112) Naik, T. *The Break-off test method. Handbook on Non-destructive Testing of Concrete*, ed. Malhotra, V.M. and Carino, N., CRC Press, Chap. 8, 1991, pp. 83–101.

(113) Stoll, U.W. Compressive strength measurement with the Stoll tork test. *Concrete International*, **7**, No. 12, Dec. 1985, 42–47.

(114) BS 1881: Part 120 *Method for determination of the compressive strength of concrete cores*. British Standards Institution, London.

(115) ASTM C42 *Standard method of obtaining and testing drilled cores and sawn beams of concrete*. American Society for Testing and Materials, Philadelphia.

(116) ACI 318 *Building code requirements for reinforced concrete*. American Concrete Institute, Detroit.

(117) Monday, J.G.L. and Dhir, R.K. Assessment of in-situ concrete quality by core testing. *Spec. Publ.* **SP 82-20**, American Concrete Institute, Detroit, 1984, 393–410.

(118) Robins, P.J. The point-load strength tests for cores. *Magazine of Concrete Research*, **32**, No. 111, June 1980, 101–111.

(119) Robins, P.J. The point-load test for tensile strength estimation of plain and fibrous concrete. *Spec. Publ.* **SP 82-16**, American Concrete Institute, Detroit, 1984, 309–325.

(120) Robins, P.J. and Austin, S.A. Core point-load test for steel fibre reinforced concrete. *Magazine of Concrete Research*, **37**, No. 133, Dec. 1985, 238–242.

(121) Clayton, N. Fluid pressure testing of concrete cylinders. *Magazine of Concrete Research*, **30**, No. 102, March 1978, 26–30.

(122) Chrisp, T.M., Walron, P. and Wood, J.G.M. Development of a non-destructive test to quantify damage in deteriorated concrete. *Magazine of Concrete Research*, **45**, No. 165, Dec. 1993, 247–253.

(123) Bartlett, F.M. and MacGregor, J.G. Effect of moisture condition on concrete core strengths. *ACI Materials Journal*, **91**, No. 3, May/June 1994, 227–236.

(124) Bartlett, F.M. and MacGregor, J.G. Effect of core length to diameter ratio on concrete core strengths. *ACI Materials Journal*, **91**, No. 4, July/Aug. 1994, 339–348.

(125) Jones, A.E.K., Clark, L.A. and Amasaki, S. The suitability of cores in predicting the behaviour of structural members suffering from ASR. *Magazine of Concrete Research*, **46**, No. 167, June 1994, 145–150.

(126) Yip, W.K. Estimating the potential strength of concrete with prior load history. *Magazine of Concrete Research*, **45**, No. 165, Dec. 1993, 301–308.

(127) Bungey, J.H. Determining concrete strength by using small diameter cores. *Magazine of Concrete Research*, **31**, No. 107, June 1979, 91–98.

(128) Bowman, S.A.W. Discussion on ref. 127, *Magazine of Concrete Research*, **32**, No. 111, June 1980, 124.

(129) Swamy, R.N. and Al-Hamed, A.H. Evaluation of small diameter core tests to determine in-situ strength of concrete. *Spec. Publ.* **SP 82-21**, American Concrete Institute, Detroit, 1984, 411–440.

(130) Schupack, M. Durability study of a 35 year old post-tensioned bridge. *Concrete International*, **16**, No. 2, Feb. 1994, 54–58.

(131) Azizinamini, A. *et al.* Old concrete slab bridges: can they carry modern traffic loads? *Concrete International*, **16**, No. 2, Feb. 1994, 64–69.

(132) Ladner, M. In-situ load testing of concrete bridges in Switzerland. *Spec. Publ.* **SP 88-4**, American Concrete Institute, Detroit, 1985, 59–80.

(133) Ladner, M. Unusual methods for deflection measurements. *Spec. Publ.* **SP 88-8**, American Concrete Institute, Detroit, 1985, 165–180.

(134) Moss, R.M. and Currie, R.J. Static load testing of building structures. *Information Paper* **IP9/89**, Building Research Establishment, Garston, 1989.

(135) Jones, D.S. and Oliver, C.W. The practical aspects of load testing. *Structural Engineer*, **56A**, No. 12, Dec. 1978, 353–356.

(136) Garas, F.K., Clarke, J.L. and Armer, G.S.T. *Structural Assessment*. Butterworths, London, 1987.

(137) Menzies, J.R., Sandberg, A.C.E. and Somerville, G. Instrumentation of structures. *Structural Engineer*, **62A**, No. 3, March 1984, 83–86.

(138) Report of working party on high alumina cement, Institution of Structural Engineers. *Structural Engineer*, **54**, No. 9, Sept. 1976, 352–361.

(139) Moss, R.M. Load testing of beam and block concrete floors. *Proc. ICE Structs & Bldgs*, **99**, May 1993, 211–223.

(140) BS 8110: Part 2 *Structural use of concrete – code of practice for special circumstances*. British Standards Institution, London.

(141) Lee, C.R., *Load testing of concrete structures, with particular reference to CP110 and experience in the HAC Investigations*. Building Research Establishment, Garston, **Doc. B507/77**, 1977.

(142) Guedelhoefer, O.C. Instrumentation and techniques of full scale testing of structures. *Experimental Methods in Concrete Structures for Practitioners*, American Concrete Institute, Detroit, 1979.

(143) Miao, B. *et al*. On-site early-age monitoring of high-performance concrete columns. *ACI Materials Journal*, **90**, No. 5, Sept./Oct. 1993, 415–420.

(144) Moss, R.M. and Matthews, S.L. In-service structural monitoring – a state of the art review. *The Structural Engineer*, **73**, No. 2, 1995, 23–31.

(145) Olaszek, P. The photogrammetric method for dynamic bridge investigations, *Proc. Struct. Faults and Repairs 93*, Eng. Technics Press, Edinburgh, **1**, 1993, pp. 89–93.

(146) Dill, M.J. and Curtis, I.L. Monitoring concrete structures using optical fibre sensors. *Concrete*, **27**, No. 5, Sept./Oct. 1993, 31–35.

(147) BS 1881: Part 206 *Recommendations for determination of strain in concrete*. British Standards Institution, London.

(148) Salah el Din, A.S. and Lovegrove, J.M. A gauge for measuring long-term cyclic strain on concrete surfaces. *Magazine of Concrete Research*, **33**, No. 115, June 1981, 123–127.

(149) Corless, R.C. and Morice, P.B. The kinematics of structural testing. *Structural Engineer*, **64B**, No. 4, Dec. 1986, 100–102.

(150) Repair of concrete damaged by reinforcement corrosion. *Tech. Rept*. **26**, Concrete Society, London, 1984.

(151) Vassie, P.R. Reinforcement corrosion and durability of concrete bridges. *Proc. ICE*, Pt. 1, **76**, Aug. 1984, 713–723.

(152) BS 1881: Part 204 *Recomendations on the use of electromagnetic covermeters*. British Standards Institution, London.

(153) Alldred, J.C. An improved method for measuring reinforcing bars of unknown diameter in concrete using a cover meter. *Proc. Non-Destructive Testing in Civil Engineering*, Brit. Inst. NDT, **2**, 1993, 767–788.

(154) Neumann, R. and Dobmann, G. *Neue magnetische Gerätetechnik bei der Beton deckung smessung*. Deutsh Gesellshaft für Zerstörungsfreie e.V., Symposium Zerstörimgsfreie Prüfung im Bauwesen, Berlin, 1991, 400–472 (in English).

(155) Carino, N.J. Characterisation of electromagnetic covermeters. *Proc. Non-Destructive Testing in Civil Engineering*, Brit. Inst. NDT, 2, 1993, 753–765.

(156) Snell, L.M., Wallace, N. and Rutledge, R.B. Locating reinforcement in concrete. *Concrete International*, **8**, No. 4, April 1986, 19–23.

(157) Alldred, J.C. Quantifying the losses in covermeter accuracy due to congestion of reinforcement. *Proc. Struct. Faults and Repairs 93*, Eng. Technics Press, Edinburgh, **2**, 1993, 125–130.

(158) ASTM C876 *Half-cell potentials of uncoated reinforcing steel in concrete*. American Society for Testing and Materials, Philadelphia.

(159) Figg, J.W. and Marsden, A.F. Development of inspection techniques–a state of the art survey of electrical potential and resistivity measurements for use above water level. Offshore Technology Report, **OTH 84 205**, HMSO, London, 1983.

(160) Vassie, P.R. *The Half-cell Potential Method of Locating Corroding Reinforcement in Concrete Structures*. TRRL Application Guide 9, 1991, 30 pp.

(161) McLaughlin, K. The detection of corroding reinforcement in concrete structures. *Brit. Jnl. NDT*, **31**, No. 12, Dec. 1989, 683.

(162) Langford, P. and Broomfield, J. Monitoring the corrosion of reinforcing steel. *Construction Repair*, **1**, No. 2, Palladian Pubs., May 1987, 32–36.

(163) Stratfull, R.F. Half cell potentials and the corrosion of steel in concrete. Highway Research Record, 1973, **433**, 12–21.

(164) Raherinaivo, A. Contrôle de la corrosion des armatures dans les structures en béton armé.

Bulletin De liason des Laboratoires des Ponts et Chaussés, **158**, Nov./Dec. 1988, 29–38 (in French).

(165) Naish, C.C. and Carney, R.F.A. Variability of potentials measured on reinforced concrete structures. *Materials Performance*, April 1988, 45–48.

(166) Arup, H. Electrochemical monitoring of the corrosion state of steel in concrete. *1st Int. Conf. Deterioration and Repair of Reinforced Concrete in the Arabian Gulf*, CIRIA/BSE, 1985, 485–493.

(167) Baker, A.F. Potential mapping techniques. *Proc. Seminar on Corrosion in Concrete*. London Press Centre, Global Corrosion Consultants, Telford, May 1986, 3.1–3.21.

(168) Millard, S.G. Durability performance of slender reinforced coastal defence units. *Spec. Publ.* **SP 109-15**, American Concrete Institute, Detroit, 1988, 339–366.

(169) Ewins, A.J. Resistivity measurements in concrete. *Brit. Jnl NDT*, **32**, No. 3, March 1990, 120–126.

(170) Millard, S.G., Harrison, J.A. and Gowers, K.R. Practical Measurement of concrete resistivity, *Brit. Jnl NDT*, **33**, No. 2, Feb. 1991, 59–63.

(171) McCarter, W.J., Forde, M.C. and Whittington, H.W. Resistivity characteristics of concrete. *Proc. ICE*, Pt. 2, **71**, March 1981, 107–117.

(172) Wilkins, N.J.M. Resistivity of concrete. Report **AERE-M3232**, UK Atomic Energy Authority, Harwell, Jan. 1982.

(173) Gowers, K.R. and Millard, S.G. The effect of steel reinforcing bars on the measurement of concrete resistivity. *Brit. Jnl NDT*, **33**, No. 11, Nov. 1991, 551–556.

(174) Millard, S.G. and Gowers, K.R. Resistivity assessment of in-situ concrete: the influence of conductive and resistive surface layers. *Proc. ICE, Structs. & Bldgs*, **94**, Nov. 1992, 389–396.

(175) Millard, S.G. Reinforced concrete resistivity measurement techniques. *Proc. ICE*, Pt. 2, March 1991, 71–88.

(176) Gowers, K.R, Millard, S.G. and Bungey, J.H. The influence of environmental conditions upon the measurement of concrete resistivity for the assessment of corrosion durability. *Proc. Non-Destructive Testing in Civil Engineering*, Brit. Inst. *NDT*, **2**, 1993, 633–657.

(177) Naish, C.C., Harker, A. and Carney, R.F.A. Concrete inspection: Interpretation of potential and resistivity measurements. *Proc. Corrosion of Reinforcement in Concrete*, Elsevier Applied Science, 1990, pp. 314–332.

(178) Lawrence, C.D. The mechanism of corrosion of reinforcement steel in concrete structures. British Cement Association, Nov. 1990, 50 p.

(179) Moore, R.W. Earth resistivity tests applied as a rapid non-destructive procedure for determining thickness of concrete pavements. Highway Research Record. Highway Research Board, Washington DC, No. 218, 1968, 49–55.

(180) John, G., Hladky, K. and Dawson, J. Recent developments in electrochemical inspection techniques for corrosion damaged structures. *Industrial Corrosion*, **11**, No. 2, Feb./Mar. 1993, 12–13.

(181) Filiu, S., Gonzalez, J.A. and Andrade, C. Electrochemical methods for on-site determination of corrosion rates of rebar. ASTM, **STP 1276**, 1995.

(182) Sehagal, A., Kho, Y.T., Osseo-Asare, K. and Pickering, H.W. Comparison of corrosion rate measuring devices for determining corrosion rate of steel-in-concrete systems. *Corrosion*, **48**, No. 10, 871–880.

(183) Elsener, B., Müller, S., Suter, M. and Böhni, H. Corrosion monitoring of steel in concrete: theory and practice. *Proc. Corrosion of Reinforcement in Concrete*, Elsevier Applied Science, 1990, pp. 348–357.

(184) Elsener, B., Wojtas, H. and Böhni, H. Inspection and monitoring of reinforced concrete structures – electrochemical techniques to detect corrosion. *Insight*, **36**, No. 7, July 1994, 502–506.

(185) Millard, S.G. Corrosion rate measurement of in-situ reinforced concrete structures. *Proc. ICE, Structs. & Bldgs*, **99**, Feb. 1993, 84–88.

(186) Dawson, J.L., John, D.G., Jafar, M.I., Hladky, K. and Sherwood, L. Electrochemical methods for inspection and monitoring of corrosion of reinforcing steel in concrete. *Proc. Corrosion of Reinforcement in Concrete*, Elsevier Applied Science, 1990, pp. 358–371.

(187) Broomfield, J.P., Rodriguez, J., Ortega, L.M. and Garcia, A.M. Field measurement of the corrosion rate of steel in concrete using a microprocessor controlled unit with a monitored

guard ring for signal confinement. *Proc. Techniques to Assess the Corrosion Activity of Steel-reinforced Concrete Structures*, ASTM, **STP 1276**, Philadelphia, 1995.

(188) Millard, S.G., Gowers, K.R. and Bungey, J.H. Galvanostatic pulse techniques: A rapid method of assessing corrosion rates of steel in concrete structures. *NACE Corrosion '95*, Paper No. 525, Orlando, USA, 16 pp.

(189) Page, C.C. and Lambert, P. Analytical and electrochemical investigations of reinforcement corrosion. TRRL Contractor Report **30**, 1986.

(190) Schiessel, P. and Raupach, M. Monitoring systems for the corrosion risk of steel in concrete structures. *Concrete International*, **14**, No. 7, 1992, 52–55.

(191) Parrott, L.J. Moisture profiles in drying concrete. *Advances in Cement Research*, **1**, No. 3, July 1988, 164–170.

(192) Petersen, C.G. and Poulsen, E. In-situ NDT methods for concrete with particular reference to strength, permeability, chloride content, and disintegration. *Proc. 1st Int. Conf. on Deterioration and Repair of Reinforced Concrete in the Arabian Gulf*, CIRIA, **1**, 1985, 495–508.

(193) Hammond, E. and Robson, T.D. Comparison of electrical properties of various cements and concretes. *The Engineer*, **199**, Jan. 1955, 78–80, 114–115.

(194) Bell, J.R., Leonards, G.A. and Dolch, W.L. Determination of moisture content of hardened concrete and its dielectric properties. *Proc. ASTM 63*, 1963, 996–1007.

(195) Jones, R. A review of the non-destructive testing of concrete. *Proc. Symp. on NDT of Concrete and Timber*, ICE, London, 1969, 1–7.

(196) Barfoot, J. Major step forward in diagnostic equipment. *Concrete*, **25**, No. 5, May 1988, 27–28.

(197) Parrott, L.J. Moisture conditioning and transport properties of concrete test specimens. *Materials and Structures*, **27**, 1994, 460–468.

(198) Parrott, L.J. A review of methods to determine the moisture conditions in concrete. British Cement Assoc., **Rep. C/7**, 1990.

(199) Newman, A.J. Improvement of the drilling method for the determination of moisture content of building materials. BRE, **CP 22/75**, 1975.

(200) Browne, J.D.I. Non-destructive testing of concrete – a survey. *Non-destructive Testing*, Feb. 1968, 159–164.

(201) Boot, A.R. and Watson, A. Applications of centimetric radio waves in non-destructive testing. In *Applications of Advanced and Nuclear Physics to Testing Materials. Spec. Tech. Publ.* **STP 373**, American Society for Testing and Materials Standards, Philadelphia, 1965, 3–24.

(202) Maser, K.R. Bridge deck condition survey using radar. Transport Research Board, Paper **91-0530**, 1991, 19 pp.

(203) Morey, R. and Kovacs, A. Detection of moisture in construction materials. US Army Cold Regions Res. & Eng. Lab., Rept. **N.77**, CRREL, 1977, 9 pp.

(204) Figg, J.W. Determining the water content of concrete panels. *Magazine of Concrete Research*, **24**, No. 79, June 1972, 94–96.

(205) Permeability of concrete – a review of testing and experience. *Tech. Rept.* **31**, Concrete Society, London, 1988.

(206) Basheer, P.A.M. A brief review of methods for measuring the permeation properties of concrete in situ. *Proc. ICE, Structs & Bldgs*, 99, Feb. 1993, 74–83.

(207) BS 1881: Part 5 *Methods of testing hardened concrete for other than strength*. British Standards Institution, London.

(208) BS 1881: Part 122 *Method for determination of water absorption*. British Standards Institution, London.

(209) Kreijger, P.C. The skin of concrete: composition and properties. *Materiaux et Constructions*, July/Aug. 1984, 275–283.

(210) Levitt, M. Non-destructive testing of concrete by the initial surface absorption method. *Proc. Symp. on NDT of Concrete and Timber*, ICE, London, 1969, 23–36.

(211) Dhir, R.K., Shaaban, I.H., Claisse, P.A. and Byars, E.A. Preconditioning in situ concrete for permeation testing – Part 1: Initial surface absorption. *Magazine of Concrete Research*, **45**, No. 163, June 1993, 113–118.

(212) Price, W.F. and Bamforth, P.B. Initial surface absorption of concrete: examination of modified test apparatus for obtaining uniaxial absorption. *Magazine of Concrete Research*, **45**, No. 162, March 1993, 17–24.

(213) Dhir, R.K., Hewlett, P.C. and Chan, Y.N. Near-surface characteristics of concrete:

assessment and development of in-situ test methods. *Magazine of Concrete Research*, **39**, No. 141, Dec. 1987, 183–195.

(214) Dhir, R.K., Hewlett, P.C., Byars, E.A. and Bai, J.P. Estimating the durability of concrete in structures. *Concrete*, Nov./Dec. 1994, 25–30.

(215) Figg, J.W. Methods of measuring air and water permeability of concrete. *Magazine of Concrete Research*, **25**, No. 85, Dec. 1973, 213–219.

(216) Cather, R., Figg, J.W., Marsden, A.F. and O'Brien, T.P. Improvements to the Figg method for determining the air permeability of concrete. *Magazine of Concrete Research*, **36**, No. 129, Dec. 1984, 241–245.

(217) Figg, J. Testing time for porous concrete. *Construction News*, **54**, 26 April, 1990.

(218) Figg, J. Concrete surface permeability: Measurement and meaning. *Chemistry & Industry*, Nov. 1989, 714–719.

(219) Hansen, A.J., Ottosen, N.S. and Petersen, C.G. Gas permeability of concrete in-situ: theory and practice. *Spec. Publ.* **SP 82-17**, American Concrete Institute, Detroit, 1984, 543–556.

(220) Basheer, P.A.M. Clam permeability tests for assessing the durability of concrete. PhD thesis, Queen's University, Belfast, 1991.

(221) Basheer, P.A.M., Long, A.E. and Montgomery, F.R. The 'Autoclam permeability system' for measuring in-situ permeation properties of concrete. *Proc. Non-Destructive Testing in Civil Engineering*, Brit. Inst. NDT, **1**, 1993, 235–260.

(222) Van der Meulen, G.J.R. and Van Dijk, J. A permeability-testing apparatus for concrete. *Magazine of Concrete Research*, **21**, No. 67, June 1969, 121–123.

(223) DIN 1048 *Test methods of concrete – impermeability to water: Part 2*. Deutscher Institut für Normung, Germany.

(224) Scales, R. Permeabiity tests on concrete and their relations to durability. *Proc. 1st Int. Conf. on Deterioration and Repair of Reinforced Concrete in the Arabian Gulf*, **1**, CIRIA, 1985, 509–527.

(225) Kelham, S. A water absorption test for concrete. *Magazine of Concrete Research*, **40**, No. 143, June 1988, 106–110.

(226) *Guidelines for the appraisal of structural components in high alumina cement concrete.* Institution of Structural Engineers, London, 1975.

(227) Schupack, M. and Schupack, D. Non-destructive field test for concrete leak tightness. *Concrete International*, **14**, No. 3, 1992, 50–54.

(228) ASTM C457 *Air void content in hardened concrete*. American Society for Testing and Materials, Philadelphia.

(229) Sadegzadeh, M., Page, C.L. and Kettle, R.J. Surface microstructure and abrasion resistance of concrete. *Cement and Concrete Research*, **17**, 1987, 581–590.

(230) Kettle, R.J. and Sadegzadeh, M. Field Investigations of abrasion resistance. *Materials and Structures*, **20**, 1987, 96–102.

(231) Sadegzadeh, M. and Kettle, R.J. Indirect and non-destructive methods for assessing abrasion resistance of concrete. *Magazine of Concrete Research*, **38**, No. 137, Dec. 1986, 183–190.

(232) Manning, D.G. and Holt, F.B. Detecting delamination in concrete bridge decks. *Concrete International*, **2**, No. 11, Nov. 1980, 34–41.

(233) Holt, F.B. and Eales, J.W. Non-destructive evaluation of pavements. *Concrete International*, **9**, No. 6, June 1987, 41–45.

(234) ASTM D4788 *Detecting delaminations in bridge decks using infrared thermography*.

(235) Godfrey, J.R. New tools help find flaws. *Civil Engineering*, **54**, No. 9, 1984, 34–41.

(236) Hillemeier, B. and Muller-Run, V. Bewehrungssuche mit der thermographie. *Beton und Stahlbeton*, **75**, 1980, 83–85.

(237) Agema Infrared Systems AB, Box 3, S-182 DANDERYD, Sweden.

(238) Stanley, C. and Balendran, R.V. Non-destructive testing of the external surfaces of concrete buildings and structures in Hong Kong using infrared thermography. *Concrete*, May/June 1994, 35–37.

(239) Ward, I.C. The use of a thermal imaging system in building and air-conditioning applications. *Insight*, **36**, No. 7, July 1994, 511–513.

(240) Cantor, T.R. Review of penetrating radar as applied to the non-destructive testing of concrete. *Spec. Publ.* **SP 82-29**, American Concrete Institute, Detroit, 1984, 581–602.

(241) Bungey, J.H. and Millard, S.G. Radar inspection of structures. *Proc. ICE, Structs & Bldgs*, **99**, May 1993, 173–186.

(242) Olver, A.D. and Cuthbert, L.G. FMCW radar for hidden object detection. *Proc. Instn Elect. Engnrs*, **135**, Part F, No. 4, Aug. 1988, 354–361.

(243) Cariou, J. Perspectives sur les applications des techniques radar aux bétons. *Bulletin de Liaison des Laboratoires des Ponts et Chaussées*, Paris, No. 163, Sept./Oct. 1989, 25–29.

(244) Bungey, J.H., Shaw, M.R., Millard, S.G. and Thomas, C. Radar testing of structural concrete. *Proc. 5th Int. Conf. Ground Penetrating Radar*, GPR '94, **1**, 305–318.

(245) Hobbs, C.P. *et al.* Radar inspection of civil engineering structures. *Proc. Non-Destructive Testing in Civil Engineering*, Brit. Inst. NDT, **1**, 1993, 79–96.

(246) Carter, C.R., Chung, T., Holt, F.B. and Manning, D.G. An automated signal processing system for the signature analysis of radar wave forms from bridge decks. *Can. Elect. Eng.*, **11**, No. 3, 1986, 128–137.

(247) ASTM D4748 *Determining the thickness of bound pavement layers using short pulse radar*.

(248) British Gas Plc. *Ground probing radar*. Publicity material, British Gas Plc, London.

(249) ERA Technology Ltd, *Surface penetrating radar*. Publicity material, Leatherhead, Surrey.

(250) GSSI Inc. *Subsurface interface radar*. Publicity material, N. Salem, New Hampshire, publicity literature.

(251) Sensors & Software Inc., Mississauga, Ontario, Canada.

(252) *Concrete International, Products & Practice*, **17**, No. 2, Feb. 1995, 60.

(253) Bungey, J.H. and Millard, S.G. Radar inspection of concrete structures. *3rd Int. Kerensky Conf., Global Trends in Structural Engineering*, Singapore, 1994, 8 pp.

(254) Cheshire, K. Police search for human remains in Gloucestershire using surface penetrating radar. *Insight*, **36**, No. 5, May 1994.

(255) Bungey, J.H., Millard, S.G. and Shaw, M.R. The influence of reinforcing steel on radar surveys of concrete structures. *Construction & Building Materials*, **8**, No. 2, 1994, 119–126.

(256) Molyneaux, T.C.K., Millard, S.G., Bungey, J.H. and Zhou, J.Q. Radar assessment of structural concrete using neural networks. *NDT and Evaluation International* (in press).

(257) ASTM D4580 *Measuring delaminations in concrete bridge decks by sounding*.

(258) *Acoustic Pile Integrity Testing Automatic Logger*. Instruction Manual, CNS Electronics Ltd, 61–63 Holmes Rd, London.

(259) Byles, R. Dutch strike a blow for integrity testing. *New Civil Engineer*, 11 June 1981, 26.

(260) Sansalone, M. Detecting delaminations in concrete bridge decks with and without asphalt overlays using an automated impact echo field system. *Proc. Non-Destructive Testing in Civil Engineering*, Brit. Inst. NDT, **2**, 1993, 807–827.

(261) Cheng, C. and Sansalone, M. Effect on impact-echo signals caused by steel reinforcing bars and voids around bars. *ACI Materials Jnl*, **90**, No. 5, Sept/Oct 1993, 421–434.

(262) Germann Instruments AS Publicity material. Emdrupvej 102, 2400 NV, Copenhagen, Denmark.

(263) A new tool for Non-Destructive measurement of pavement thickness. *SHRP Focus*, July 1994, 1–4.

(264) Casas, J.R. An experimental study on the use of dynamic tests for surveillance of concrete structures, RILEM. *Materials & Structures*, 1994, **27**, 588–595.

(265) Carino, N.J. and Sansalone, M. Detection of voids in grouted ducts using the impact-echo method. *ACI Materials Jnl.*, **89**, No. 3, May/June 1992, 296–303.

(266) Pratt, D. and Sansalone, M. Impact-echo signal interpretation using artificial intelligence. *ACI Materials Jnl.*, **89**, No. 2, Mar/Apr 1992, 178–187.

(267) Olson, L.D. and Wright, C.C. Non-destructive testing for repair and rehabilitation. *Concrete International*, **12**, No. 3, Mar. 1990, 58–64.

(268) Nazarian, S. and Desai, M.R. Automated surface wave method: Field testing. *ASCE Jnl Geotech. Eng.*, **119**, No. 7, July 1993, 1094–1111.

(269) Maguire, J.R. and Severn, R.T. Assessing the dynamic properties of prototype structures by hammer testing. *Proc. ICE*, Pt. 2, **83**, Dec. 1987, 769–784.

(270) Williams, C. Vibration testing of large structures to determine structural characteristics. *Proc. Struct. Faults and Repair 87*, Engineering Technics Press, Edinburgh, 31–35.

(271) Stain, R.T. Integrity testing. *Civil Engineering*, April 1982, 54–59 and May 1982, 71–73.

(272) BS 1881: Part 205 *Recommendations for radiography of concrete*. British Standards Institution, London.

(273) Forrester, J.A. Gamma radiography of concrete. *Proc. Symp. on NDT of Concrete and Timber*, ICE London, 1969, 9–13.
(274) Pullen, D. and Clayton, R. The Radiography of Swathling bridge. *Atom*, No. 301, Nov. 1981, 283–288.
(275) Parker, D. X-rated video. *New Civil Engineer*, 21st April, 1994, 8–9.
(276) Carter, P. Problems with post-tensioned bridges. *QT News*, **53**, Mar. 1995, Harwell A-EA Technology, p. 4.
(277) Kear, P and Leeming, M. Radiographic inspection of post-tensioned concrete bridges. *Insight*, **36**, No. 7, July 1994, 507–510.
(278) Smith, E.E. and Whiffin, A.C. Density measurement of concrete slabs using gamma radiation. *The Engineer*, **194**, Aug. 1952, London, 278–281.
(279) de Hass, E. Radio-isotope techniques in concrete research. *ACI Journal, Proceedings*, **50**, No. 10, June 1954, 890–891.
(280) Simpson, J.W. A non-destructive method of measuring concrete density using backscattered gamma radiation. *Building Science*, **3**, No. 1, Aug. 1968, Pergamon, Oxford, 21–29.
(281) Honig, A. Radiometric determination of the density of fresh shielding concrete in-situ. *Spec. Publ.* **SP 82-30**, American Concrete Institute, Detroit, 1984, 603–618.
(282) Martz, H.E., Schneberk, D.J., Robertson, G.P. and Monteiro, P.J.M Computerised tomography analysis of reinforced concrete. *ACI Materials Jnl*, **90**, No. 3, May/June 1993, 259–264.
(283) Hawkins, N.M., Kobayashi, A.S. and Fourney, M.E. Use of holographic and acoustic emission techniques to detect structural damage in concrete members. *Experimental Methods in Concrete Structures for Practitioners*, American Concrete Institute, Detroit, 1979.
(284) Nielsen, J. and Griffin, D.F. Acoustic emission of plain concrete. *Journal of Testing and Evaluation (JTEVA) USA*, **5**, No. 6, Nov. '77, 476–483.
(285) Lim, M.K. and Koo, T.K. Acoustic emission from reinforced concrete beams. *Magazine of Concrete Research*, **41**, No. 149, Dec. 1989, 229–234.
(286) Dunegan/ENDEVCO *3000 Series Acoustic Emission Instrumentation System*. ENDEVCO (UK Division), Mebourn, Royston, Herts.
(287) Mindess, S. Acoustic emission and ultrasonic pulse velocity of concrete. *Int. Journal of Cement Composites and Lightweight Concrete*, **4**, No. 3, Construction Press, Aug. 1982, pp. 173–179.
(288) Titus, R.N.K., Reddy, D.V., Dunn, S.E. and Hartt, W.H. Acoustic emission crack monitoring and prediction of remaining life of corroding reinforced concrete beams. *Proc. 4th European Conf. on NDT*, Vol. 2, Pergamon, Oxford, 1988, pp. 1031–1040.
(289) Rossi, P., Godart, N., Robert, J.L., Gervais, J.P. and Bruhat, D. Investigations of the basic creep of concrete by acoustic emission. *Materials and Structures*, **27**, 1994, 510–514.
(290) Jenkins, D.R. and Steputat, C.C. Acoustic emission monitoring of damage initiation and development in structurally reinforced concrete beams. *Proc. Struct. Faults & Repairs 93*, Eng. Technics Press, Edinburgh, **3**, 1993, 79–87.
(291) Redner, A.S. Use of photoelastic and Moiré fringe techniques. *Experimental Methods in Concrete Structures for Practitioners*, American Concrete Institute, Detroit, 1979.
(292) ASTM C1074 *Estimating concrete strength by the maturity method*. American Society for Testing and Materials, Philadelphia.
(293) Pearson, R.I. Maturity meter speeds post-tensioning of structural concrete frame. *Concrete International*, **9**, No. 4, April 1987, 63–64.
(294) Carino, N.J. The maturity method. *CRC Handbook on Non-Destructive Testing of Concrete*. CRC Press, ed. Malhotra, V.M. and Carino, N.J., Chapt. 5, 1991, pp. 101–146.
(295) BS DD92 *Method of temperature matched curing of concrete specimens*. British Standards Institution, London.
(296) Harrison, T.A. Temperature matched curing. *Proc. Symp. on Developments in Testing Concrete for Durability*, Concrete Society, London, Sept. 1985, 25–34.
(297) Cannon, R.P. Temperature matched curing: its development, application and future role in concrete practice and research. *Concrete*, **20**, No. 7, July 1986, 27–30; **20**, No. 10, Oct. 1986, 28–32; and **21**, No. 2, Feb. 1987, 33–34.
(298) Pye, P.W. and Warlow, W.J. A method of assessing the soundness of some dense floor screeds. *Current Paper* **CP72/78**, Building Research Establishment, Garston, 1978.
(299) Pye, P.W. BRE Screed tester: classification of screeds, sampling and acceptance limits. *Information Paper* **IP11/84**, Building Research Establishment, Garston, 1984.

(300) Chung, H.W. and Law, K.S. Assessing fire damage of concrete by the ultrasonic pulse technique. *Cement, Concrete, and Aggregates, CCAGDP*, **7**, No. 2, American Society for Testing and Materials, Philadelphia, Winter 1985, 84–88.

(301) Evaluation and repair of fire damage to concrete. *Spec. Publ.* **SP-92**, American Concrete Institute, Detroit, 1986.

(302) BS 1881: Part 124 *Methods of testing concrete: analysis of hardened concrete.* British Standards Institution, London.

(303) Figg, J.W. and Bowden, S.R. *The Analysis of Concrete*, HMSO, London, 1971.

(304) Concrete Society Analysis of hardened concrete, *Technical Report*, **32**, 1989.

(305) ASTM C85 *Cement content of hardened Portland cement concrete.* American Society for Testing and Materials, Philadelphia.

(306) Grantham, M.C. Determination of slag and pulverised fuel ash in hardened concrete – the method of last resort revisited. **STP 1253**, Amer. Soc. for Testing and Materials, 1995 (in press).

(307) Brown, A.W. A tentative method for the determination of the original water/cement ratio of hardened concrete. *Journal of Applied Chemistry*, 7 Oct. 1957, 565–572.

(308) Mayfield, B. The quantitative evaluation of the water/cement ratio using fluorescence microscopy. *Magazine of Concrete Research*, **42**, No. 150, March 1990, 45–49.

(309) Neville, A.M. *Properties of Concrete.* Pitman, 1981, 512.

(310) Power, T.O. and Hammersley, G.P. Practical concrete petrography. *Concrete*, **12**, No. 8, Aug. 1978, 27–30.

(311) French, W.J. Concrete petrography: a review. *Quarterly Jnl of Eng. Geology*, Geological Soc., **24**, 1991, 17–48.

(312) Roberts, H.M. and Jaffrey, S.A.M.T. Rapid chemical test for the detection of high alumina cement concrete. *Information Sheet* **IS 15/74**, Building Research Establishment, Garston, 1974.

(313) ASTM 856 *The petrographic examination of hardened concrete.* American Society for Testing and Materials, Philadephia.

(314) BS 812: Part I *Methods for sampling and testing of mineral aggregates, sands and fillers.* British Standards Institution, London.

(315) Determination of chloride and cement content in hardened Portand cement concrete. *Information Sheet* **IS 13/77**, Building Research Establishment, Garston, 1977.

(316) Simpified method for the detection and determination of chloride in hardened concrete. *Information Sheet* **IS 12/77**, Building Research Establishment, Garston, 1977.

(317) Figg, J.W. Methods to determine chloride concentrations in in-situ concrete. TRRL Contractor Report **32**, 1986.

(318) 'Hach' Chloride test kit, Camlab Ltd., Nuffield Rd., Cambridge.

(319) 'Quantab' Chloride filtrators, Miles Laboratories Ltd., Stoke Court, Stoke Poges, Slough, Bucks.

(320) BS 8110: Part 1 *Structural use of concrete: code of practice for design and construction.* British Standards Institution, London.

(321) Kay, E.A., Fookes, P.G. and Pollock, D.J. Deterioration related to chloride ingress. *Concrete*, **15**, No. 11, Nov. 1981, 22–28.

(322) Roberts, M.H. Carbonation of concrete made with dense natural aggregates **IP6/81**, Building Research Establishment, Garston, 1981.

(323) Theophilus, J. Uncertainties in assessing the durability of concrete. *Civil Engineering*, Aug. 1986, 10–13.

(324) Sims, I. The assessment of concrete for carbonation. *Concrete*, **28**, No. 6, Nov./Dec. 1994, 33–38.

(325) Watkins, R.A.M. and Pitt-Jones, A.A. Carbonation: a durability model related to site data. *Proc. ICE, Structs. and Bldgs*, **99**, May 1993, 155–166.

(326) Spectronics Corporation. Detector keeps bridges from vanishing. *Concrete International*, **16**, No. 7, July 1994, 78.

(327) Natesaiyer, K., Stark, D. and Hover, K.C. Gel fluorescence reveals reaction product traces. *Concrete International*, **13**, No. 1, Jan. 1991, 25–28.

(328) Polivka, M., Kelly, J.W. and Best, C.H. A physical method for determining the composition of hardened concrete. *Spec. Tech. Publ.* **STP 205**, 1958, American Society for Testing and Materials, Philadelphia, 135–152.

(329) Markestad, A. Changing over from standard microscopic technique to an image analysing

microscope for measuring quality parameters of concrete. *Proc. Int. Conf. on Concrete Structures*, Trondheim, Tapir, 1978, pp. 57–67.

(330) Poulsen, E. *Preparation of samples for microscopic investigation.* Progress Report M1, Committee on alkali reactions in concrete, Danish National Institute of Building Research and Academy of Technical Sciences, Copenhagen, 1958.

(331) Placido, F. Thermoluminescence test for fire damaged concrete. *Magazine of Concrete Research*, **32**, No. 111, June 1980, 112–116.

(332) Chew, M.Y.L. Effect of heat exposure duration on the thermoluminescence of concrete. *ACI Materials Journal*, **90**, July/Aug. 1994, 319–322.

(333) Midgley, H.G. The mineralogy of set high alumina cement. *Transactions of the British Ceramic Society*, **66**, No. 4, June 1967, 161–187, *Current Paper* **CP 19/68**, Building Research Establishment, Garston.

(334) Midgley, H.G. and Midgley, A. The conversion of high alumina cement. *Magazine of Concrete Research*, **27**, No. 91, June 1975, 59–77; *Current Paper* **CP 72/75**, Builing Research Establishment, Garston.

(335) Ramachandran, V.S. *Calcium Chloride in Concrete*, Applied Science (London), 1976, pp. 22–38.

(336) Willard, H.H., Merritt, L.L. and Dean, J.A. *Instrumental Methods of Anaysis*, 5th edn., Van Nostrand, London, 1974.

Index